Technological Innovations for Sustainable Industrialization

Xavier Poshiwa · Vennan M. Sibanda ·
Boitumelo Ramatsetse
Editors

Technological Innovations for Sustainable Industrialization

Springer

Editors
Xavier Poshiwa
Gary Magadzire School of Agriculture and Engineering
Great Zimbabwe University
Masvingo, Zimbabwe

Boitumelo Ramatsetse
Department of Mechanical and Mechatronics Engineering
Stellenbosch University (SU)
Western Cape, South Africa

Vennan M. Sibanda
Innovation and Technology Transfer, Innovation and Business Development
National University of Science and Technology (NUST)
Bulawayo, Zimbabwe

ISBN 978-981-96-7994-2 ISBN 978-981-96-7995-9 (eBook)
https://doi.org/10.1007/978-981-96-7995-9

This Springer imprint is published by the registered company Springer Nature Singapore Pte Ltd.
The registered company address is: 152 Beach Road, #21-01/04 Gateway East, Singapore 189721, Singapore

Foreword

In an era defined by rapid technological advancements and an urgent need for sustainability, the intersection of innovation and industrialization has never been more critical. The global economy faces an unprecedented challenge: how to achieve industrial growth without compromising the environment. This book, *Technological Innovations for Sustainable Industrialization*, explores this vital convergence, offering a roadmap for industries, policymakers and researchers striving for a greener future.

The pursuit of sustainable industrialization is a defining challenge of our time, requiring innovative technological solutions that balance economic growth with environmental responsibility. This collection of research, *Technological Innovations for Sustainable Industrialization*, brings together groundbreaking studies exploring novel approaches to industrial advancements across various sectors, from **nanotechnology and precision agriculture to infrastructure development and renewable energy integration**.

One of the highlights of this work is the emphasis on **bio-synthesis and eco-friendly technologies**, such as the use of **Iraqi propolis for the production of magnetic iron oxide nanoparticles**, which has shown promising results in treating iron deficiency anaemia. This innovation underscores the potential of **biotechnology and nanomedicine** in addressing pressing healthcare challenges through sustainable means.

Similarly, the role of **precision agriculture** as an enabler of **sustainable farming in developing countries** is explored in depth, emphasizing how mechanization, research and local innovation can transform smallholder farming. The research also delves into **anaerobic treatment of yeast processing effluent**, showcasing the impact of temperature on wastewater treatment efficiency—an essential aspect of industrial sustainability.

Beyond scientific and technological advancements, this book highlights the **importance of research-driven policies, entrepreneurship and academia-industry collaboration** in shaping a sustainable industrial future. The insights presented here offer valuable perspectives for **policymakers, researchers, industry leaders and entrepreneurs**, providing a roadmap for harnessing technology in a way that fosters both economic development and environmental sustainability.

As we move forward, the integration of cutting-edge technologies with sustainable practices will be crucial in addressing global challenges. This book serves as both a **testament to human ingenuity and a call to action** for stakeholders across industries to innovate responsibly and collaboratively for a better future.

The industrial revolutions of the past transformed societies, fueling economic progress but often at the expense of ecological balance. Today, as we stand on the brink of the Fourth Industrial Revolution, we have an opportunity—and a responsibility—to redefine the foundations of industrialization. Emerging technologies such as artificial intelligence, the Internet of Things, renewable energy, circular economy models, and sustainable manufacturing practices provide innovative solutions that can drive economic prosperity while minimizing environmental impact.

This book brings together insights from experts across various fields, illustrating how cutting-edge technologies can revolutionize industries, reduce carbon footprints, optimize resource use and promote inclusive economic growth. From smart factories and green supply chains to energy-efficient production techniques, the discussions here highlight practical solutions and future possibilities.

As you journey through these pages, I encourage you to reflect on the role each of us plays in shaping a sustainable industrial future. Whether you are an entrepreneur, a researcher, a policymaker or an industrial leader, the knowledge within this book will inspire action, innovation and transformation.

Sustainable industrialization is not just a choice, it is an imperative. The path forward requires collaboration, creativity and commitment to harness technology for the benefit of both industry and the planet. May this book serve as a guide and a catalyst for meaningful change.

October 2025

Thomas Varghese
Ex-Business Head, Textiles, Acrylic
Fibre and Overseas Spinning
Aditya Birla Group
Mumbai, India

Preface

Sustainable industrialization stands as one of the most formidable challenges of the twenty-first century, lying at the intersection of technological advancement, environmental stewardship and social equity. As the world grapples with escalating pressures from climate change, resource depletion and social inequality, the urgency to adopt sustainable industrial practices has never been greater. Against this backdrop, the book *Technological Innovations for Sustainable Industrialization* offers a comprehensive exploration of technology's dynamic and multifaceted role in shaping a future where sustainable industrial growth aligns with environmental protection, societal well-being and economic prosperity. By focusing on practical applications, real-world case studies and cutting-edge innovations, this work intends to fill the critical gap in the existing literature, offering actionable insights and strategies for industries seeking to achieve sustainability goals.

This volume is the product of collaboration among a diverse group of leading experts, each contributing their knowledge and experience to address the most pressing challenges of sustainable industrialization. It provides an in-depth examination of how innovations can drive industrial growth while minimizing adverse environmental and societal impacts. The chapters in the book span a wide array of topics, including clean energy technologies, sustainable manufacturing practices, digital transformation in industries and the integration of advanced technologies into industrial systems. Beyond presenting cutting-edge research and emerging trends, the book offers practical applications and detailed case studies that demonstrate innovation's transformative potential in fostering sustainability across sectors.

The purpose of this publication is to synthesize a broad spectrum of perspectives, ensuring a holistic and inclusive exploration of how technological advancements can pave the way for a more sustainable industrial future. It not only highlights the challenges hindering sustainable industrial practices but also identifies actionable solutions adaptable to different contexts and scales. The contributions provide valuable insights into how sectors such as energy, manufacturing, agriculture and transportation can harness advanced technologies to build more sustainable and resilient systems. Furthermore, the book's interdisciplinary approach and diverse global perspectives make it an essential resource for stakeholders, including researchers, technology developers, policymakers, industry leaders and students—committed to advancing sustainable industrialization.

At its core, this book recognizes that sustainable industrialization is not a static goal but an ongoing dynamic process requiring continuous adaptation, innovation and collaboration. As industries strive to meet the demands of a growing global population, technological innovation must align with principles of sustainability, resilience and social equity. Only then can we ensure that the benefits of industrial progress are equitably shared, securing long-term prosperity without compromising the well-being of future generations. When directed towards sustainability, technological innovation can propel us towards a future that is not only productive but also equitable and resilient—an industrial paradigm that serves both people and the planet.

This book is the result of collective effort, and we extend our deepest gratitude to all who contributed to its realization. We sincerely thank the authors whose expertise and dedication have made this publication possible. Their insightful work provides invaluable resources for addressing some of the most critical challenges facing industries worldwide. We also acknowledge the reviewers, whose constructive feedback refined and enhanced each chapter, ensuring the book delivers accurate, timely and relevant insights.

Our special thanks goes to Dr. Amitava Bandopadhyay, Director General of the NAM S&T Centre, along with the concerned staff members, for their steadfast support in facilitating the technical and administrative aspects of this publication. Their contributions were instrumental in bringing this work to fruition.

We hope that this book, published by the NAM S&T Centre, New Delhi through Springer Nature, Singapore, will benefit readers from diverse backgrounds by offering practical guidance and actionable insights. In particular, the book addresses key aspects of the UN Sustainable Development Goals (SDGs), with a strong focus on *SDG 9: Industry, Innovation, and Infrastructure*. By highlighting innovative solutions that drive sustainable industrialization, the publication aims to contribute to the global efforts to build resilient infrastructure, foster innovation, and promote inclusive and sustainable industrialization. As the world navigates the complexities of sustainable industrialization, collective action and shared expertise remain our most powerful tools for overcoming challenges and seizing opportunities.

We trust that the ideas presented here will accelerate the global transition towards a more sustainable, equitable, and resilient industrial future for all.

Masvingo, Zimbabwe — Xavier Poshiwa, Ph.D.
Bulawayo, Zimbabwe — Vennan M. Sibanda, Ph.D.
Stellenbosch, South Africa — Boitumelo Ramatsetse, Ph.D.

Introduction

In the era marked by rapid technological advancements, global urbanization, and escalating environmental challenges, the need for sustainable industrialization has never been more urgent. However, for industrialization to be sustainable, it is essential to leverage cutting-edge innovations and at the same time, address the dynamic challenges of climate change, resource depletion and urbanization. As we seek to transform the process of industrial development, whether in energy, transportation, agriculture or manufacturing—innovation will be a key driver of success.

The concept of *Technological Innovations for Sustainable Industrialization* underscores the transformative potential of emerging technologies and creative solutions in reshaping the way we manage, design and build our industry and infrastructure. Innovations—ranging from renewable energy technologies and advanced materials to digital tools and automation—are pivotal in enhancing the resilience, efficiency and environmental sustainability of the development process. In particular, these innovations can play a crucial role in driving industrialization in developing economies, where rapid urbanization and population growth often outpace the ability to build adequate infrastructure. By integrating cutting-edge research and innovation into infrastructure planning and industrial development, we can ensure that growth is not only sustainable but also inclusive and equitable.

To address these issues, the Centre for Science and Technology of the Non-Aligned and Other Developing Countries (NAM S&T Centre) presents this publication titled *Technological Innovations for Sustainable Industrialization* which brings together contributions from scientists and experts from seven countries, viz., India, Iraq, Mauritius, Myanmar, Nigeria, South Africa and Zimbabwe.

Through its 14 chapters, the book aims to provide both a detailed overview and actionable insights into how innovations can reshape the future of infrastructure and industrialization. By highlighting pioneering research and case studies, it offers valuable lessons for policymakers, engineers, entrepreneurs and researchers working towards building more resilient, inclusive and sustainable industrialization.

The book is structured into three key parts, each addressing different aspects of leveraging innovations for infrastructure and industrial development. First Part focuses on foundational technologies that support sustainable industrial development, including innovative solutions for addressing public health challenges, agriculture technologies and wastewater treatment systems. Second Part explores the relationship between research, innovation and entrepreneurship, highlighting how innovation-driven ecosystems, educational reforms, and the development of IT skills for business students can drive industrial growth and entrepreneurship, particularly in developing economies. Third Part examines the application of advanced technologies, with case studies on agro-processing, space technology, digitalization, energy transition challenges and revitalization of traditional sectors like textiles, showcasing how these technologies contribute to sustainable industrialization.

I extend my heartfelt gratitude to all the contributors for sharing their expertise, which has made this compilation possible. I specially thank the editorial team—Dr. Xavier Poshiwa [Zimbabwe], Dr. Vennan M. Sibanda [Zimbabwe] and Dr. Boitumelo Ramatsetse [South Africa] for their relentless efforts in reviewing and refining the chapters, ensuring that the information presented is both accurate and impactful. I am also grateful to some other experts for their help in reviewing a few papers included in the book.

I am also immensely grateful to Mr. Thomas Varghese, Ex-Business Head—Textiles, Acrylic Fibre and Overseas Spinning, Aditya Birla Group, Mumbai, India, for kindly agreeing to our request to write the Foreword of this book in spite of his very busy schedule.

I am thankful to Dr. Loyola D'Silva, Executive Editor, Springer Nature, Singapore, for his kind support and guidance towards bringing out this publication and Mrs. Kavitha Palanisamy, Production Editor, Springer Nature, Singapore, for managing all the technical and administrative responsibilities for the publication process.

I would like to take this opportunity to express my sincere gratitude to Mr. Madhusudan Bandyopadhyay, Senior Adviser at the NAM S&T Centre, for his invaluable advice and guidance throughout the planning and execution of this important publication. I also thank Ms. Nidhi Utreja, Programme Officer, NAM S&T Centre, for her tireless efforts in coordinating the work on this publication. I also wish to acknowledge the hard work and support of the entire NAM S&T Centre team, especially Mr. Pankaj Buttan, Data Processing Manager, and Ms. Abhirami Ramdas, Programme Officer—whose contributions have played a vital role in bringing this publication to completion.

I hope that the book will inspire new collaborations, spark further research initiatives and ignite actions towards application of technological innovations for sustainable industrialization and achieving the global sustainable development goals.

Amitava Bandopadhyay, Ph.D.
Director General
NAM S&T Centre
New Delhi, India

Contents

Part I
Foundational Technologies

Chapter 1
Bio-Synthesis of Iron Oxide Nanoparticles Using Iraqi Propolis as a Reducing Agent for the Treatment of Iron Deficiency Anemia

Ryaidh Sh. Al-Hussain and Majida A. J. Al-Qayim

Abstract The present study aimed to restore homeostasis of iron regulatory mechanism in the treatment of iron deficiency anemia by using a synthesized form of iron oxide nanoparticles (IONPs), which are characterized by their small size, low toxicity, cost-effectiveness, and high efficacy. The study included two parts. First part involved the biosynthesis of IONPs using aqueous extract of Iraqi propolis as a reducing agent. The synthesized magnetite nanoparticles were characterized by color changes, UV–visible absorption spectroscopy, Fourier transform infrared spectroscopy (FTIR), X-ray diffraction (XRD), and scanning electron microscopy (SEM) techniques. Results of this part revealed that the best temperature was 80 °C, pH 3.8, and the volume ratio was 1:2 (propolis to iron precursors). The involvement of propolis phenolic groups in iron oxide nanoparticles synthesis was evident from the results of the FTIR analysis. The XRD results at 2θ revealed that the average particle size of magnetic nanoparticles formed from a mixture of $FeCl_3$ and $FeCl_2$ was found to be in the range of 31.71–46.20 nm, and SEM analysis of these nanoparticles showed crystalline structured particles. The second part of the study included two experiments: The first aimed to induce iron deficiency anemia (IDA) in a population of experimental rats, while the second experiment investigated the efficacy of propolis-coated IONPs (Pr-IONPs) in restoring iron homeostasis and iron regulatory mechanisms that were disrupted in IDA. Results of this study showed that magnetized iron oxide nanoparticles are essentially a new method characterized by a one-step synthesis, lower cost, and healthier profile, due to the synthesis of small sized particles. Furthermore, the use of propolis–IONPs improved the regulatory mechanisms for iron homeostasis in iron-deficient anemia. A single I/p dosage of

R. Sh. Al-Hussain (✉)
Ministry of Science and Technology to the Scientific Research Authority, Ministry of Higher Education and Scientific Research, Baghdad, Iraq
e-mail: Ryaidh1969@gmail.com

M. A. J. Al-Qayim
Department of Physiology, College of Veterinary Medicine, University of Baghdad, Baghdad, Iraq

X. Poshiwa et al. (eds.), *Technological Innovations for Sustainable Industrialization*,
https://doi.org/10.1007/978-981-96-7995-9_1

60 mg/kg of Pr-IONPs could be considered an appropriate treatment for iron deficiency anemia when compared with iron dextran, which requires several injections. From the results of the present study, it can be concluded that the bio–production of Magnetic Iron Oxide Nanoparticles were successfully achieved through an easy, non-toxic, low cost, and eco-friendly method.

Keywords Green synthesis · Iron oxide nanoparticles · Anemia

Introduction

Iron deficiency anemia is a condition that describes insufficient oxygen delivery to tissues. It is a global health problem, given its impact on psychological well-being, behavior, physical development, and work performance (WHO, 2017). The most common form of iron deficiency occurs when iron stores are inadequate for the body's demands. This may happen due to reduced iron intake, reduced iron absorption by the gut, and increased iron loss (Auerbach and Deloughery, 2016). The prevalence of iron deficiency anemia (IDA) in humans ranges from 2 to 10% in developed countries, with the highest rate in women of reproductive age (Camaschella, 2015; Shander et al., 2014). The most frequent clinical manifestation of IDA is a reduction in the oxygen-carrying capacity of red blood cells, resulting in many adverse symptoms associated with reduced health-related quality of life. Recently, IDA has been associated strongly with pica in pregnant women (Roy et al., 2018). Reduced bioavailabilities of dietary iron and increased iron requirement, as well as obesity, are significantly associated with iron deficiency (Zhao et al., 2015), chronic disease (Madu & Ughasoro, 2017), and chronic blood loss, especially in dogs and cats (Dinaz et al., 2012). The biological importance of iron, including cellular respiration (Saporito et al., 2017), energy production, DNA synthesis, and cell proliferation (Puig et al., 2017), make it essential for sustaining life. Treatment of IDA is essential to improve health and performance. Iron may be supplemented through various routes, some administered orally, such as ferrous sulfate and others via I/V injection, such as iron gluconate, iron saccharate, and low molecular weight iron dextrin (Liudmyla et al., 2013).

However, oral iron supplements have important disadvantages. Oral iron is not well tolerated by all patients; it causes side effects such as constipation, diarrhea, nausea and stomach pain (Lopez et al., 2015). Another disadvantage is that iron absorption through the oral route is limited by intestinal conditions (Bregman et al., 2013). Furthermore, intestinal iron absorption is dependent on erythropoiesis. When iron stores are sufficient, the hepcidin limits iron absorption (Ganz & Nemeth, 2012).

Intravenous (IV) iron supplements are directly taken up by liver cells, liver macrophages, and erythrocytes. The ability of macrophages to uptake IV iron varies depending on the specific iron formulation (Kari et al., 2015).

Nanotechnology, the science involved in the design, synthesis and application of materials with a nanoscale size ranging from 1 to 100 nm (Liu et al., 2010), has opened new avenues in medicine. The application of iron oxide nanoparticles as a drug formulation to treat anemia represents a novel drug delivery system. These nanoparticles can be mixed with other compounds such as folic acid, nicotinic acid (vitamin B3), and ascorbic acid (vitamin C). This new formula has been developed for the rapid and efficient treatment of iron-deficiency anemia, one of the most life-threatening conditions (Fabio et al., 2005).

Although the biological synthesis of iron oxide by green plants has been extensively studied, the green production of Fe-based bimetallic nanoparticles using various plants, such as black and green tea, has gained attention. In addition, plants that can reduce Fe also exhibit antioxidant activity and provide beneficial effects to the organism.

Propolis, a resinous substance collected by honey bees, serves as a beehive shield and is characterized by its brownish-green color. The biochemical composition of propolis is highly diverse, consisting mainly of a mixture of polyphenols, flavonoids, aglycones, phenolic acids and their esters, phenolic aldehydes and ketones, terpenes, sterols, vitamins, and amino acids. Due to its rich composition to maximize the efficacy and safety of biosynthesized IONPs, we propose a novel method using propolis for their production.

The aim of this study was the biosynthesis and characterization of a novel method for IONPs synthesis using Iraqi propolis and to evaluate the efficacy of the propolis-derived IONPs in treating iron deficiency anemia in rats.

Materials and Methods

Biosynthesis of Iron Oxide Nanoparticles (IONPs)

The synthesis of nanoparticles by biological methods may lead to development of non-toxic, non-polluting, and environmentally acceptable nanoparticles that are biologically safe. For these reasons, propolis was used as a source for the biosynthesis of iron oxide nanoparticles.

Crude propolis was obtained from a local market as Iraqi propolis. After cutting the resin blocks of propolis, the extract was prepared using the maceration method. Briefly, 50 gm of propolis was placed in a glass beaker, mixed with 500 ml of distilled water, and continuously stirred at 35 °C for 3 days. Then, the suspension was filtered using 0.1 μm filter paper and the supernatant was cooled to room temperature and stored at 4 °C until use.

Propolis was considered a good substance for reducing iron salts, such as their oxides. Since propolis is a resinous material characterized by a high content of polyphenolic compounds, it was used for the production of IONPs under different conditions, including varying pH levels, different ratios between iron salts and propolis extract, and iron salt combinations. The synthesis was carried out by mixing

0.1 M $FeCl_3$ and $FeCl_2$ with propolis extract in a volume ratio of 1: 2, with stirring at 80 °C and pH 3.8.

Characterization of Iron Oxide Nanoparticles

The characteristics of synthesized iron oxide nanoparticles were analyzed by UV–Visible absorption spectroscopy, Fourier transform infrared analysis (FTIR), X-ray diffraction (XRD), and scanning electron microscope (SEM).

Evaluation Effect IONPs Synthesized by Propolis on Iron Deficiency Anemic Rats

Induction and Assessment of Iron Deficiency Anemia

The first experiment of the second part was designed to induce iron deficiency anemia (IDA) in growing female rats: 50 female rats (21 days of age) were divided into two groups randomly: a control group fed an iron adequate diet, and an iron-deficient anemic (IDA) group including forty (40) female rats, which were fed an iron-deficient diet. The diet was formulated according to the AIN-93G rodent diet, with some modifications based on ingredients available in the local markets.

Evaluation of the Effect of IONPs on Iron Deficiency Anemic Rats

The second experiment was designed to study the beneficial effects of propolis mediated synthesis of magnetite nanoparticles on iron homeostasis in iron-deficient anemic rats, compared to iron dextran. Ten (10) healthy female rats and forty (40) iron-deficient female rats handled as follows:

1st group (IAD): Control group; ten female rats fed on iron-adequate diet (IAD) with iron 40 mg/ kg diet.
2nd group (IDA): Iron-deficient anemic group; fed an iron- deficient diet (IDD) with iron = 5–10 mg/kg diet.
3rd group (IDA2X): Iron-deficient anemic group; fed IDD with iron = 5–10 mg/ kg diet and treated intraperitoneally (i/p) with two doses of IONPs (30 mg/kg body weight) at ten day intervals.
4th group (IDAX2): Iron-deficient anemic group; fed IDD with iron = 5–10 mg/ kg diet and treated with a single dose of IONPs (60 mg/kg body weight).

5th group (IDAD): Iron-deficient anemic group; fed IDD with iron = 5–10 mg/kg diet and treated intraperitoneally (I/p) with iron dextran (30 mg/kg body weight three days).

The experiment lasted for 30 days. During the experiment, all anemic rats remained on the same feeding regimes except for the control group. Feeding and water supply were provided ad libitum. Body weight for all experimental animals was recorded at the beginning and end of the experiment for calculating body weight changes. At the end of the experiment, blood and tissue samples were collected.

Statistical Analysis

The data of the present study were analyzed using SPSS Version 2010. The SPSS statistical software program was used for all data analysis by applying one and two way analysis of variance (ANOVA). The least significant difference (L.S.D.) test was used to compare means. An alpha value of $P < 0.05$ was regarded as statistically significant. Data were expressed as Mean ± SE.

Results and Discussion

Preparation and Characterization of Effective Compounds in Iraqi Propolis Extract

Diagnostics of effective compounds in Iraqi propolis extract were performed using analytical instruments, focusing on flavonoid and polyphenol compounds, absorbance determination, and identification of the active materials present. It is the most suitable and reliable analytical method for identifying and quantifying propolis and green tea active biomolecules. The relatively polar nature of polyphenols, with multiple hydroxyl groups in their structure, combined with UV absorbance of the aromatic rings and soft ionization techniques compatible with chromatography, make high-performance liquid chromatography (HPLC) one of the most valuable and repeatable methods for the characterization and determination of the active compounds in green tea and propolis (Ali, 2020).

The results of an evaluation of active materials (polyphenols and flavonoids) in Iraqi propolis extract included: Rutin, Quercetin, Kaempferide, Isorhamnetin, Catechin, and Coumarins, which were evaluated by their chromatographic performance and retention times (Ali, 2020). These concentrations were calculated according to the Eq. (1.1) (Hammadi et al., 2020)

$$\text{Concentration} = \frac{\text{area of the sample} \times \text{Concentration of standard}}{\text{area of standard}}. \quad (1.1)$$

Table 1.1 The concentrations of active materials in Iraqi propolis extract

Propolis extract	
Concentrations ppm(μg/ml)	Compounds
133.41	Rutin
36.16	Quercetin
327.25	Kaempferide
254.55	Isorhamnetin
154.16	Catechin
190.68	Coumarins

Table 1.1 showed the polyphenols and flavonoids and their concentrations. Figures 1.1 and 1.2 showed the polyphenols and flavonoids in Iraqi propolis and standard chromatograms. The qualitative analysis of Iraqi Propolis extract demonstrated numerous active ingredients, such as flavonoids and phenolic acids. The flavonoids and phenolics in Iraqi Propolis extract were abundant and had reducing functional groups. These functional groups may play a role in reducing zinc oxide to zinc oxide nanoparticles (ZnO NPs). Biomolecules can act as biocatalysts in the fabrication of nanomaterials. They can also serve as natural nanoparticles stabilizers. Plant-mediated or natural product-based synthesis reduces the use of harmful chemicals and results in a more environmentally friendly production of NPs.

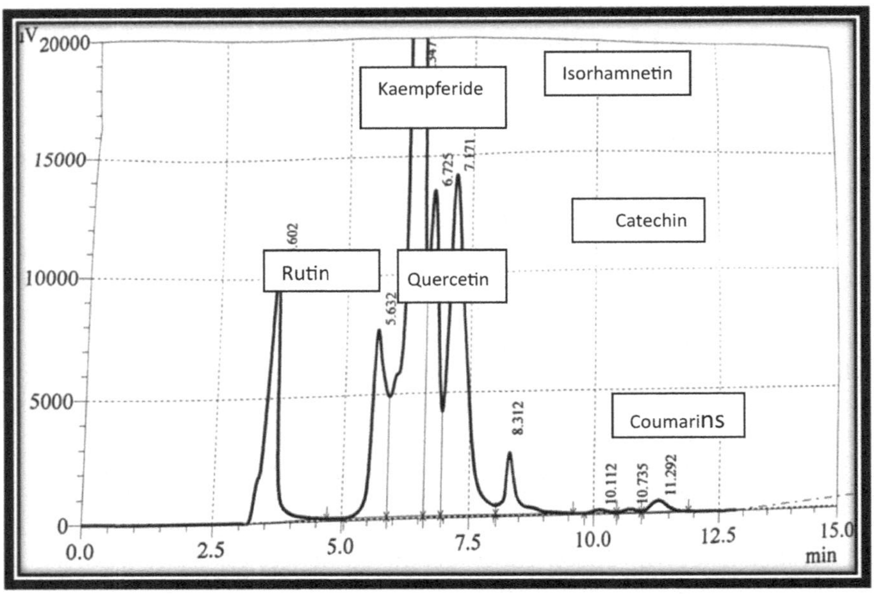

Fig. 1.1 Chromatogram of active material in Iraqi propolis

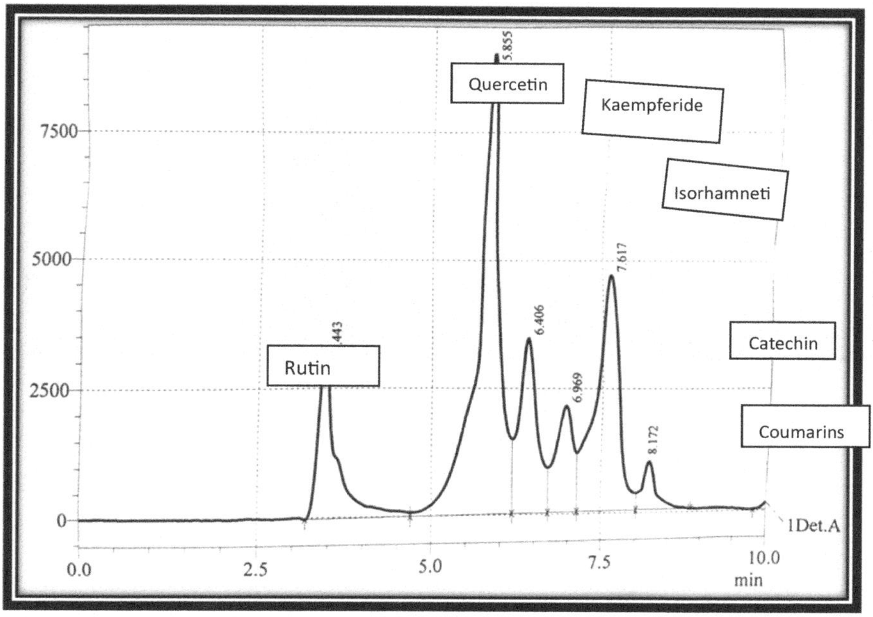

Fig. 1.2 Chromatogram of standard polyphenols

Biosynthesis and Characterization of Magnetic Iron Oxide Nanoparticles

The biosynthesis of magnetic oxide nanoparticles was prepared by bio-reduction of ferric and ferrous salts using propolis water extract in the present study and was clearly documented by color changes, FTIR, XRD and SEM. The color changes of the synthesized nanoparticles using propolis extract are shown in Fig. 1.3. There was a change in color within minutes after mixing the propolis extract with iron salts, shifting from yellow to black, along with an increase in the density and intensity of the precipitate. Ferric chloride undergoes hydrolysis, forming ferric hydroxide and releasing protons (H^+). In the presence of aldehyde groups, magnetic nanoparticles are formed, and the aldehyde is oxidized to form the corresponding acid (Cahyana et al., 2017; Kumar et al., 2014).

Fourier Transform Infrared (FTIR) Spectroscopy

The analysis of the FTIR spectrum for the prepared compound is shown in Fig. 1.4. Each region of wavelength indicates the presence of bonded molecules. Results revealed that the prepared iron oxide nanoparticles using propolis compounds were

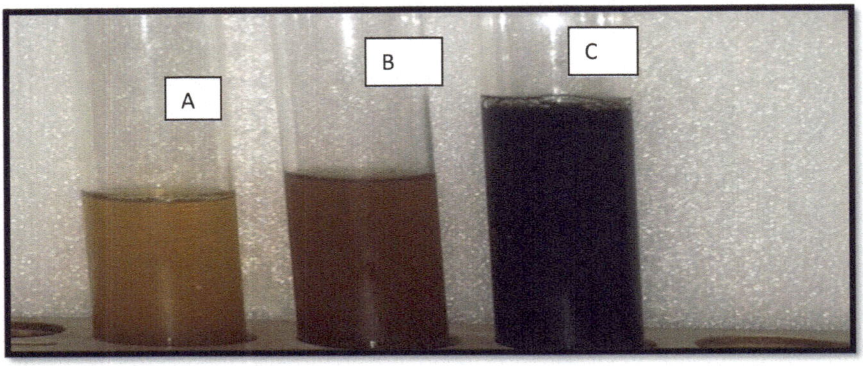

Fig. 1.3 Changes in color and precipitate density immediately after mixing iron salt mixture with propolis extract in a 1:2 ratio. A: Iron salt mixture B: Propolis extract C: The iron salt: propolis mixture

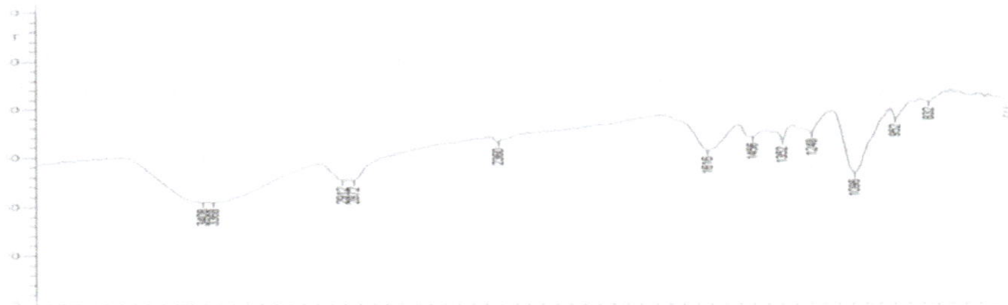

Fig. 1.4 FTIR spectrum of iron oxide nanoparticles synthesis by propolis extract

rich in O–H, C–C, N–H, and COOH groups, with IR absorption bands in the range of 3400–1500 cm. In addition, they showed absorption in the fingerprint region below 750–480 which is specific for iron oxide bounds. FTIR analysis was conducted to evaluate the functional groups in the propolis extract and to predict their role in the formation of IONPs. These molecules were OH, CH, COO, NH, C=O, which are involved in the formation of Fe–O bonds. This technique was used to establish the metallic nature of the particles and provides information about the atom is positioning based on peak intensities (Rajiv et al., 2013; Rajeshwari et al., 2014).

X-Ray Diffraction Analysis

The analysis of the XRD spectrum included the 2θ values of the differentiated peaks shown in Fig. 1.5. The crystallite particle sizes were calculated using Schrerre's

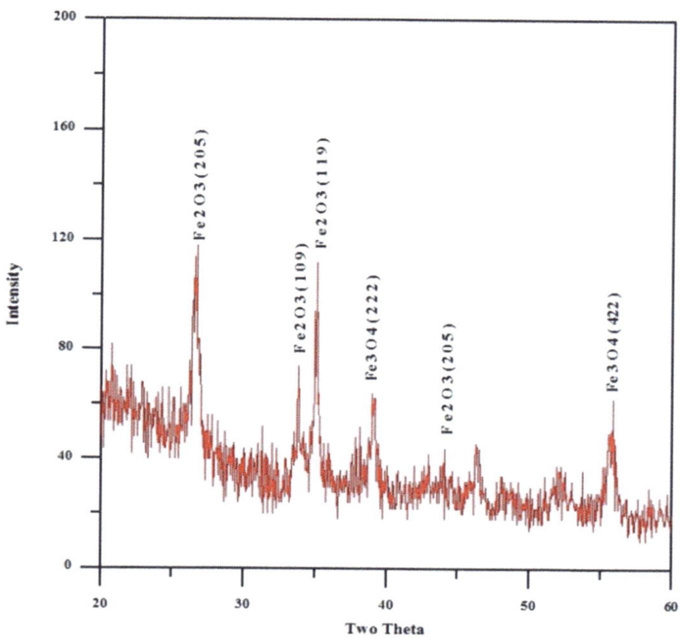

Fig. 1.5 XRD spectrum of iron oxide nanoparticles synthesized by propolis extract

formula for the different types of iron oxide nanoparticles prepared using propolis extract. The results showed sharp diffraction peaks at 2θ angles, depending on the type of iron oxide formed.

Phase identification and crystalline structure of the nanoparticles were determined by identifying strong diffraction peaks with 2θ values of 26.618°, 33.8515°, 35.05°, 39.07°, 46.375°, and 55.7°, corresponding to the hkl values of 205, 109, 119, 222, and 422 crystal planes, respectively. PCPDFWIN-PDF #25-1402 and PDF #26-1136 were used for reference, with a wavelength of 1.54056 Å.

XRD analysis was an effective characterization technique to confirm the crystal structure of the synthesized magnetite nanoparticles from propolis extract. A strong peak with 2θ values corresponding to the formed type of iron oxide was observed. The XRD results demonstrated that the nanoparticles were crystalline, especially in the IONPs produced from a mixture of iron salts, which exhibited a cubic shape and were identified as Fe_3O_4 and Fe_2O_3 (Shojaee & Shahri, 2016). The strong diffraction peaks at specific 2θ values indicated the biosynthesis of iron oxide nanoparticles, including magnetite nanoparticles (Fe_3O_4 and Fe_2O_3). These results were consistent with previous studies on the green synthesis of magnetite nanoparticles (Sathishkumar et al., 2018).

Scanning Electron Microscopy Analysis

The present results from the analysis of the SEM micrographs, represented in Fig. 1.6, showed the size and morphology of magnetic nanoparticles synthesized using propolis extract. These images clearly revealed a distinct cubic crystalline shape, with a particle size range of chrysalis size 36.7–42 nm in low pH propolis solution. Particles formed at high pH showed agglomeration of iron oxide nanoparticles. The nanoparticles prepared using propolis extracts exhibited different shapes and sizes, depending on the formulation processes. The magnetite nanoparticles produced in PrIONP1 were cubic in shape, with varying particles sizes. This outcome can be explained by the fact that the polyphenol concentration in the propolis extract plays a critical role in determining the final structure and particle size of the biosynthesized magnetite nanoparticles, as demonstrated in green synthesis approaches (Mahdavi et al., 2013).

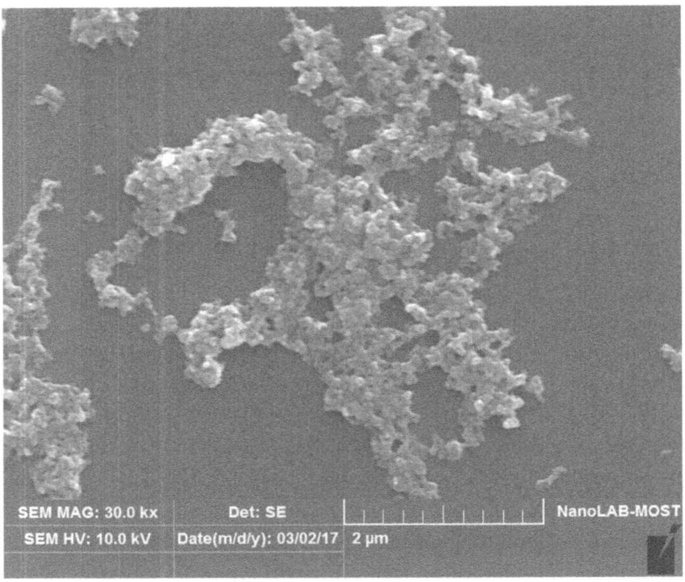

Fig. 1.6 SEM spectrum of iron oxide nanoparticles synthesized by propolis extract

Evaluation the Effect IONPs on Iron Deficiency Anemic Rats

Evaluation of Bone Marrow (Myeloid/Erythroid Ratio)

The results of the bone marrow assessment in the experimental rats are shown in Fig. 1.7. Anemic rats exhibited a significant reduction in the myeloid-to-erythroid (M/E) ratio. At the same time, anemic rats treated with prepared magnetic iron oxide nanoparticles and iron dextran showed a significant increase in the M/E ratio. The values of the M/E ratio in IDA, IDA2X and IDAX2 Groups were analyzed. Analysis of the results in Fig. 1.7 revealed that PrIONPs were more efficient in restoring the M/E ratio to normal levels than iron dextran. The decreased ratio between myeloid and erythroid cells in IDA rats may result from increased bone marrow production of red blood cells, which leads to a reduction in myeloid cells.

Iron deficiency causes hypoxia in the bone marrow, as in any other tissue in the body, suggesting that erythropoiesis is increased in response to the anemia. As a result, the bone marrow exhibits erythroid hyperplasia and a decrease in the myeloid-to-erythroid ratio (Sahar & Soltan, 2013). Other studies have reported that iron acts on mature cell lines in one way and, interestingly, in an inverse way on immature cell lines (Torres et al., 2017).

Iron Status

The total iron significantly ($P < 0.05$) decreased in IDA group. There were no significant differences ($P > 0.05$) in iron concentration in IDA2X, IDAX2, and IDAD (see Table 1.2). While comparing the remaining groups, there was a significant ($P < 0.05$) increase toward near normal or semi-normal value.

The level of total iron binding capacity (TIBC) significantly increased ($P < 0.05$) in the IAD group compared to the IDA group. On the other hand, the levels of TIBC significantly decreased ($P < 0.05$) in the IDA2X, IDAX2, and IDAD groups, approaching the levels observed in the IAD group.

The level of unsaturated iron-binding capacity (UBIC) significantly increased ($P < 0.05$) in the IDA group compared to the IAD group. The levels of UBIC significantly decreased ($P < 0.05$) in the IDA2X, IDAX^2 and IDAD near the IAD group. The levels of ferritin (fr) statistically decreased ($P < 0.05$) in the IDA group compared to the IAD group. The levels of ferritin significantly increased ($P < 0.05$) in the IDA2X, IDAX2, and IDAD groups, approaching those in the IAD group. Systemic reduction of serum iron in anemic rats after 90 days was accompanied by decreased levels of serum ferritin and increased total iron-binding capacity. Under normal conditions, plasma iron is bound to transferrin. Transferrin can bind large amounts of iron without reaching saturation. In iron-deficient anemic rats, the high TIBC and UBIC are attributed to elevated transferrin levels in response to reduced iron availability, as the liver responds to low circulating iron by increasing transferrin production to meet iron demand (Yamanishi et al., 2003). The decrease in total body

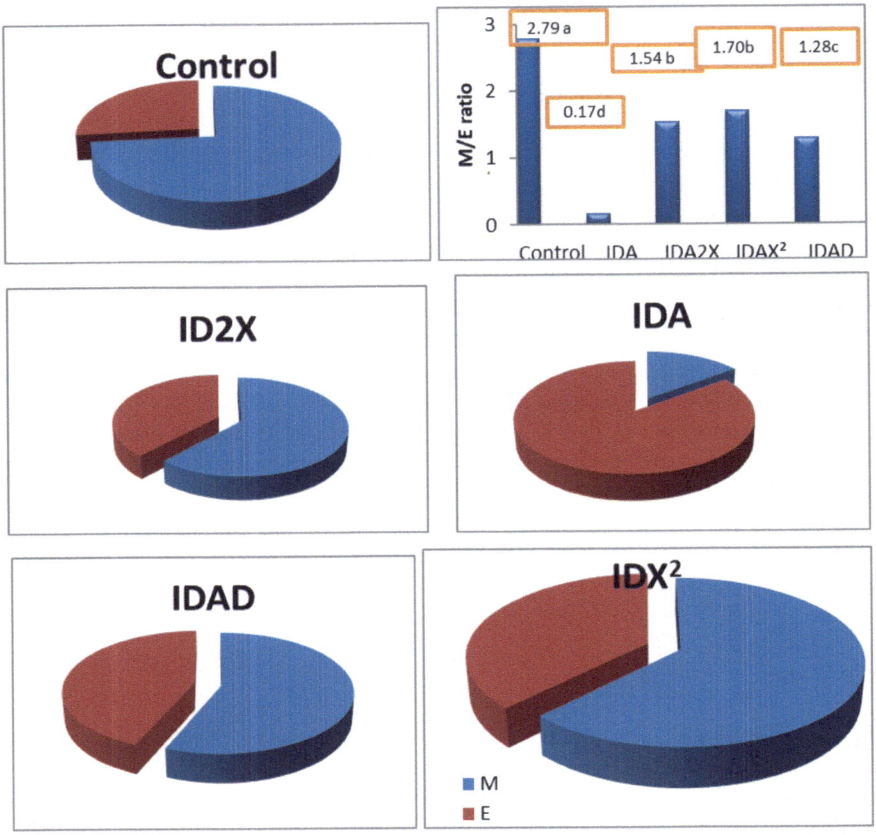

Fig. 1.7 Role of propolis–IONPs in restoring myeloid/erythroid ratio rehomeostasis in the bone marrow of iron deficiency anemic female rats, compared with iron dextran treatment. Mean ± SE. L.S.D. = 0.236

Table 1.2 Role of propolis-IONPS on iron (μg/dl), TBIC (μg/dl), UBIC (μg/dl), and ferritin (μg/l) in iron deficiency anemic female rats compared with iron dextran, M ± SE

GROUP	Iron (μg/dl)	TIBC (μg/dl)	UBIC (μg/dl)	Ferritin (μg/l)
IAD	302.75 ± 21.74^{a}	643.54 ± 7.43^{b}	340.79 ± 18.38^{b}	1.32 ± 0.0024^{cb}
IDA	110.74 ± 13.01^{b}	852.04 ± 20.63^{a}	738.29 ± 24.54^{a}	1.27 ± 0.0043^{d}
IDA2X	339.95 ± 32.65^{a}	637.89 ± 40.25^{b}	297.72 ± 27.78^{b}	1.33 ± 0.0088^{ab}
IDAX2	339.53 ± 61.51^{a}	625.09 ± 43.51^{b}	293.99 ± 24.38^{b}	1.34 ± 0.0097^{a}
IDAD	240.94 ± 19.79^{a}	583.59 ± 27.78^{b}	342.65 ± 26.91^{b}	1.31 ± 0.0023^{c}
LSD	99.929	89.933	71.724	0.0186

iron is easily diagnosed based on reduced levels of serum ferritin and transferrin saturation. Iron deficiency leads to metabolic disorders. The beneficial effects of Pr-IONPs in iron-deficient rats were clarified by the significant elevation of total serum iron and systemic reduction in transferring binding capacity, indicating that dissolved iron ions were bound to transferrin. This elevation in circulating iron resulted in increased intracellular iron levels.

Magnetic iron oxide nanoparticles uptake is likely mediated by endocytotic uptake of the particles, as was recently demonstrated by other types of nanoparticles (Busch et al., 2011).

Transferrin Saturation (TfS %)

The levels of transferrin saturation (TfS) statistically decreased ($P < 0.05$) in the IDA group compared with the IAD group. The levels of TfS significantly increased ($P < 0.05$) in the IDA2X, IDAX2 IDAX^2 and IDAD groups (Fig. 1.8) as compared to IDA. Plasma transferrin is known as a central player in iron metabolism, responsible for circulating iron in a soluble, non-toxic form and delivering it to the erythron and other tissues. Iron delivery from transferrin to cells occur via receptor mediated endocytosis through transferrin receptor1 (TfR1) (Mayle et al., 2012).

Transferrin is a member of the iron regulatory protein (IRP) network, which plays a role in the regulation of cellular iron metabolism. Furthermore, transferrin captures iron released into the plasma, mainly from intestinal enterocytes or macrophages. Normally, there is daily variation in transferrin saturation (Elsayed et al., 2016). The TfR1 receptor can be released into circulation and is referred to as soluble transferrin receptor (sTfR). Therefore, sTfR is a useful marker of iron-deficient erythropoiesis, reflecting both absent iron stores and restricted iron supply due to anemia in chronic disease.

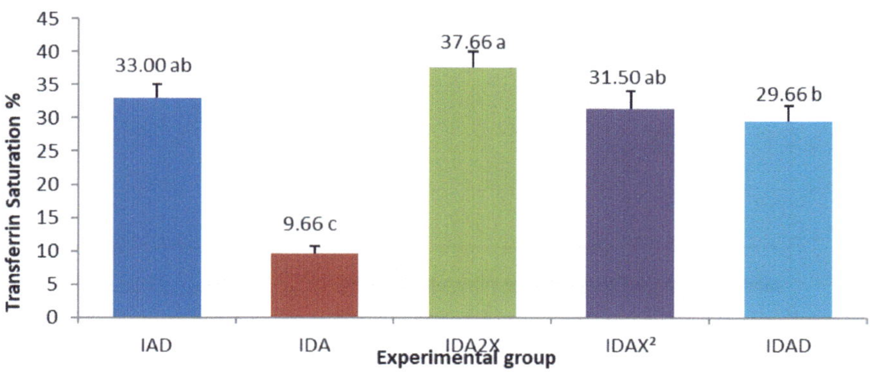

Fig. 1.8 Role of propolis-IONPs in restoring transferrin saturation (Tf%) rehomeostasis in iron deficiency anemic female rats, compared to iron dextran treatment. L.S. D. = 6.33

Gene Expression of Hepcidin by Hepatocytes (Hamp)

Changes to gene expression of hepcidin in hepatic tissue of anemic and treated anemic rats across different experimental groups were clarified in Figs. 1.9 and 1.10. The results revealed that iron deficiency anemia had a significant ($P < 0.05$) suppressive effect on the expression of hepcidin. There was a statistically significant decrease in hepcidin expression, reaching a minimum of 0.13 ± 0.03 after 90 days of the experimental period, compared with the control group 1.03 ± 0.20. Results indicated that hepcidin transcription was positively regulated by the treatment of anemic rats with propolis-IONPs, at 30 and 60 mg/kg body weight, as well as with iron dextran. The highest fold induction of hepcidin transcription was observed in the IDAX2 group (6.29 ± 0.76), compared to IDA2X (2.02 ± 0.21) and IDAD (3.19 ± 0.22). The connection between hepcidin and iron metabolism was first recorded by Pigeon et al. (2001) through the investigation of hepatic responses to iron overload.

The main role of hepcidin is its interference with iron regulatory mechanisms and homeostasis (Daher et al., 2017), through direct action on the cellular iron exporter, ferroportin. Under normal conditions, gastrointestinal tract iron absorption is considered a dynamic process, which can be suppressed in order to maintain stable extracellular iron level. Studies have pointed out that increased hepcidin inhibits iron absorption in the small intestine, leading to reduced release of recycled iron from macrophages. According to these observations, hepcidin is considered a key component of iron homeostasis that acts as a negative regulator of iron metabolism. Systemic iron levels are controlled by the hepcidin protein (Hamp), as hepcidin regulates the entry of iron into circulation and is secreted by the liver (Ganz & Nemeth, 2012). Hamp induces degradation of ferroportin protein during elevated iron level in human

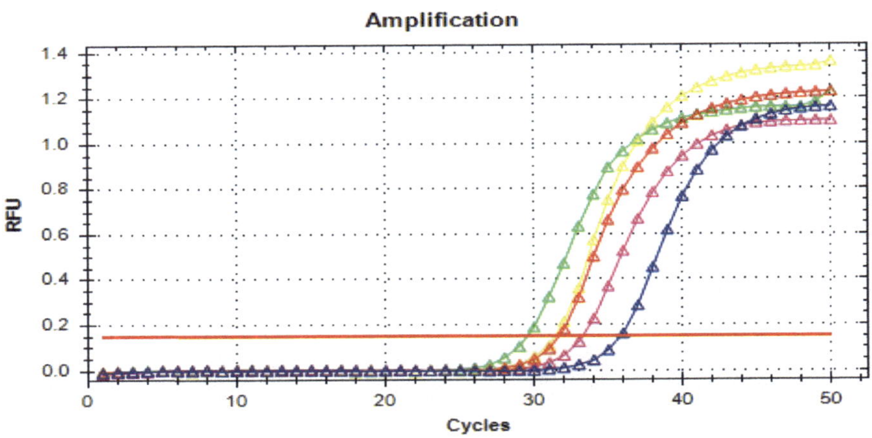

Fig. 1.9 Role of propolis–IONPs in the re-establishment of homeostasis in hepatic hepcidin expression in iron deficiency anemic female rats. Pink plot: IAD group, Blue plot: IDA group, Red plot: IDA2X group, Green plot: IDAX2 group, Yellow plot: IDAD group

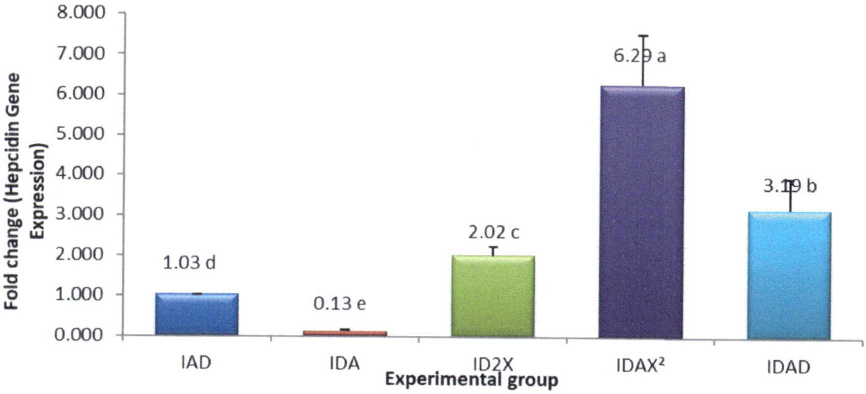

Fig. 1.10 Role of propolis–IONPs in restoring hepatic hepcidin expression rehomeostasis in iron deficiency anemic female rats. L.S.D. = 1.75

body. Inflammatory cytokines induce hepcidin overproduction, which causes endocytosis and proteolysis of ferroportin (FPN), favoring iron trapping in absorbing enterocytes, resulting in iron deficiency anemia (IDA) (Shah et al., 2018). Serum hepcidin levels are reduced in iron-deficient anemic subjects (Litton et al., 2016).

Conclusion

From the results of the present study, it can be concluded that the bio–production of magnetic iron oxide nanoparticles was successfully achieved in an easy, non-toxic, low cost, eco-friendly, and less time-consuming way using Iraqi propolis extracts rich in polyphenol compounds as both reducing and capping agents. These extracts were used to reduce iron salts and prepare magnetite and maghemite iron oxide nanoparticles (IONPs) that possess superparamagnetic properties and small particles size, contributing to iron regulation in iron deficiency anemic rats.

In vivo results indicated that I/P injection of magnetic iron oxide nanoparticles achieved great success in treating IDA in the experimental animal model, achieving complete recovery of anemic rats from the anemic state to a normal hemoglobin (Hb) level within 30 days. The injection of a single dose of 60 mg/kg body weight of iron oxide nanoparticles proved to be more effective compared with the iron dextran drug.

References

Ali, U. (2020). *Green synthesis, characterization & therapeutic evaluation of ZnO nanoparticles prepared using extract of Nigella sativa seeds*. PhD dissertation, Capital University.

Auerbach, M., & Deloughery, T. (2016). Single dose intravenous iron for iron deficiency: A new paradigm. *American Society of Hematology, 2016*, 57–66.

Bregman, D. B., Morris, D., Koch, T. A., He, A., & Goodnough, L. T. (2013). Hepcidin levels predict nonresponsiveness to oral iron therapy in patients with iron deficiency anemia. *American Journal of Hematology, 88*(2), 97–101.

Busch, W., Bastian, S., Trahorsch, U., Iwe, M., Kuhnel, D., Meisner, T., Springer, A., Gelinsky, M., Richter, V., & Ikonomidou, C. (2011). Internalisation of engineered nanoparticles into mammalian cells in vitro: Influence of cell type and particle properties. *Journal of Nanoparticle Research, 13*, 293–310.

Cahyana, A. H., Pratiwi, D., & Ardiansah, B. (2017). Citrate-capped superparamagnetic iron oxide (Fe_3O_4-CA) nanocatalyst for synthesis of pyrimidine derivative compound as antioxidative agent. *Materials Science and Engineering, 1886*, 1–4.

Camaschella, M. D. (2015). Iron deficiency anemia. *New England Journal of Medicine, 372*, 1832–1843.

Daher, R., Hana, M., & Zoubida, K. (2017). Iron metabolism and the role of the iron-regulating hormone hepcidin in health and disease. *Presse Medicale (Paris, France: 1983), 2017*, 1–7.

Dinaz, Z., Naigamwalla, J., Webb, A., & Urs, G. (2012). Iron deficiency anemia. *Canadian Veterinary Journal, 53*, 250–256.

Elsayed, M. E., Sharif, M. U., & Stack, A. G. (2016). Transferrin saturation: A body iron biomarker. *Advances in Clinical Chemistry, 75*, 71–77.

Fabio, S., Stephane, M., Sebastien, V., Catherine, D., Danielle, J., Jeril, D., Johan, H., Etienne, D., Paolo, C., & Patrick, C. (2005). Folate conjugated iron oxide nanoparticles for solid tumor targeting as potential specific magnetic hypertheria mediators: Synthesis, physicochemical characterizati. *Bioconjugate Chemistry, 16*(5), 1181–1188.

Ganz, T., & Nemeth, E. (2012). Hepcidin and iron homeostasis. *Biochimica Et Biophysica Acta, 1823*(9), 1434–1443.

Hammadi, A. H., Habeeb, S. A., Al-Jibouri, L. F., & Hussien, F. H. (2020). Synthesis, characterization and biological activity of zinc oxide nanoparticles (ZnO NPs). *Systematic Reviews in Pharmacy, 11*(5), 431–439.

Kari, A., Duck, J., & Connor, R. (2015). Iron uptake and transport across physiological barriers. *BioMetals, 29*, 573–591.

Kumar, B., Kumari, S., Luis, C., & Alexis, D. (2014). Biogenic synthesis of iron oxide nanoparticles for 2-arylbenzimidazole fabrication. *Journal of Saudi Chemical Society, 2014*, 1–6.

Litton, E., Baker, S., Erber, W. N., Farmer, S., Ferrier, J., French, C., Gummer, J., Hawkins, D., Higgins, A., Hofmann, A., De Keulenaer, B., McMorrow, J., Olynyk, J. K., Richard, T., Towler, S., Trengove, R., & Webb, S. (2016). Intravenous iron or placebo for anaemia in intensive care: the iron man multicenter randomized blinded trial: a randomized trial of iv iron in critical illness. *Intensive Care Medicine, 42*, 1715–1722.

Liu, C., Uwe, B., Flemming, B., & Zhong, L. (2010). Preparation and characterization of nanomaterials for sustainable energy production. *Nano Focus, 4*(10), 5517–5526.

Liudmyla, R., Andrii, D., & Tamara, G. (2013). Iron nanoparticles effectiveness for new antianemic preparations development. *Nano Conference, 10*, 16–18.

Lopez, A., Patrice, C., Iain, C. M., & Laurent, P. B. (2015). Iron deficiency anaemia. *Seminar, 2015*, 1–10.

Madu, A., & Ughasoro, J. (2017). Anaemia of chronic disease: An in-depth Review. *Medical Principles and Practice, 26*, 1–9.

Mahdavi, M., Farideh, N., Mansor, B. A., & Rosfarizan, M. (2013). Green biosynthesis and characterization of magnetic iron oxide (Fe_3O_4) nanoparticles using seaweed (*Sargassum muticum*) aqueous extract. *Molecules, 18*(5), 5954–5964.

Mayle, K. M., Le, A. M., & Kamei, D. T. (2012). The intracellular trafficking pathway of transferrin. *Biochimica et Biophysica Acta, 1820*(3), 264–281.

Pigeon, C., Gennady, I., Brice, C., Patricia, L., Bruno, T., Pierre, B., & Olivier, L. (2001). A new mouse liver-specific gene, encoding a protein homologous to human antimicrobial peptide hepcidin, is overexpressed during iron overload. *Journal of Biological Chemistry, 276*, 7811–7891.

Puig, S., Lucia, R. A., Antonia, M. R., & Maria, T. M. (2017). The elemental role of iron in DNA synthesis and repair. *Metallomics, 2017*, 1–59.

Rajeshwari, S., Pattanathu, K. S. M., Rahman, R. P., Narendhran, S., & Venckatesh, R. (2014). Biosynthesis and characterization of *Acalypha indica* mediated copper oxide nanoparticles and evaluation of its antimicrobial and anticancer activity. *Spectrochimica Acta A: Molecular and Biomolecular Spectroscopy, 129*, 255–258.

Rajiv, P., Rajeshwari, S., & Venckatesh, R. (2013). Bio-fabrication of zinc oxide nanoparticles using leaf extract of *Parthenium hysterophorus* L. and its size-dependent antifungal activity against plant fungal pathogens. *Spectrochimica Acta Part A: Molecular and Biomolecular Spectroscopy, 112*, 384–385.

Roy, L., Soiza, A., Donaldson, I. C., & Phyo, K. M. (2018). The pale evidence for treatment of iron-deficiency anaemia in older people. *Therapeutic Advances in Drug Safety, 2018*, 1–3.

Sahar, S., & Soltan, A. (2013). The protective effect of soybean, sesame, lentils, pumpkin seeds and molasses on iron deficiency anemia in rats. *World Applied Sciences Journal, 23*(6), 795–807.

Saporito, M. C., Musacco-Sebio, R., Acosta, J. M., Bajicoff, S., Paredes-Fleitas, P., Reynoso, S., Boveris, A., & Repetto, M. G. (2017). Copper (II) and iron (III) ions inhibit respiration and increase free radical-mediated phospholipid peroxidation in rat liver mitochondria: Effect of antioxidants. *Journal of Inorganic Biochemistry, 172*, 94–99.

Sathishkumar, G., Logeshwaran, V., Sarathbabu, S., Pradeep, K., Jeyaraj, M., Rajkuberan, C., Senthilkumar, N., & Sivaramakrishnan, S. (2018). Green synthesis of magnetic Fe3O4 nanoparticles using *Couroupita guianensis* Aubl. fruit extract for their antibacterial and cytotoxicity activities. *Artificial Cells Nanomedicine and Biotechnology, 46*(3), 1–10.

Shah, A., Wray, K., James, T., Shine, B., Morovat, R., Stanworth, S., Mckechnie, S., Kirkbride, R., Griffith, D. M., Walsh, T. S., Drakesmith, H., & Roy, N. (2018). Serum hepcidin potentially identifies iron deficiency in survivors of critical illness at the time of hospital discharge. *British Journal of Haematology, 2018*, 1–3.

Shander, A., Goodnough, L. T., Javidroozi, M., Auerbach, M., Carson, J., Ershler, W. B., Ghiglione, M., Glaspy, J., & Lew, I. (2014). Iron deficiency anemia bridging the knowledge and practice gap. *Transfusion Medicine Reviews, 28*(3), 156–166.

Shojaee, S., & Shahri, M. (2016). Green synthesis and characterization of iron oxide magnetic nanoparticles using Shanghai White tea (*Camelia sinensis*) aqueous extract. *Journal of Chemical and Pharmaceutical Research, 8*(5), 138–143.

Torres, M. M., Cruz, F. A. C. S., Érica, P. S., Daiane, P., Cayuela, A. A., Adriane, J. M., Arleana, B. P., & Almeida, F. (2017). Relation between anaemia and bone marrow features and serum erythropoietin in dogs with chronic kidney disease. *Pesquisa Veterinária Brasileira, 37*(6), 598–602.

World Health Organization. (WHO). *Iron deficiency anaemia. Assessment, prevention and control. A guide for programme managers.* World Health Organization: Geneva, Switzerland (2017).

Yamanishi, H., Iyama, S., Yamaguchi, Y., Kanakura, Y., & Iwatani, Y. (2003). Total iron binding capacity calculated from serum transferrin concentration or serum iron concentration and unsaturated iron binding capacity. *Clinical Chemistry, 49*(1), 175–178.

Zhao, L., Zhang, X., Shen, Y., Fang, X., Wang, Y., & Wang, F. (2015). Obesity and iron deficiency: A quantitative meta-analysis. *Obesity Reviews, 16*(12), 1081–1093.

Chapter 2
Precision Agriculture as a Sustainable Farming Enabler in Developing Countries

Pradeep Rajan

Abstract Conventional agricultural practices result in negative impacts on the soil, water, and environment due to over and imprecise application of farm inputs. Managing the small farm holdings as a homogeneous block was considered as a kind of precision agriculture; however, field studies established the existence of spatial and temporal variation in soil and plant properties which can be further managed to optimize the yield potential. Precision agriculture (PAg) comprises information, technologies, and a management strategy to maximize the yield potential with optimized farm inputs in a sustainable way. Except for Brazil, PAg in most of the developing countries is in the nascent stage due to different adoption and implementation issues. It may be seen from the recent developments that recommendations regarding irrigation scheduling, fertilizer application, and pesticide application are gaining popularity in India, and these must be wholly assisted with field-level technologies like advanced farm equipment suitable for small and marginal farmers. Custom hiring and Farming as a Service (FaaS) are possible ways to demonstrate, implement and sustain the advantages of the PAg in the longer run. This paper addresses the status, constraints, and future prospects of PAg in developing countries. The use of drones in India got a boost due to the supportive government policy and several startups are offering services in mapping the farm field for plant growth monitoring and pesticide application. PAg technologies, popular in developed countries, may not be a solution because developing countries have their inherent constraints of low capital, fragmented and small field size, education level of farmers, and immediate benefits of the technology. Precision agriculture with the newly evolving cyber-physical systems (CPS) can transform conventional agricultural systems into sustainable farming.

Keywords Precision agriculture · Small farm holdings · Developing countries · Sustainable farming

P. Rajan (✉)
CSIR—CMERI Centre of Excellence for Farm Machinery, Ludhiana, Punjab, India
e-mail: prajan.cmeri@csir.res.in

X. Poshiwa et al. (eds.), *Technological Innovations for Sustainable Industrialization*,
https://doi.org/10.1007/978-981-96-7995-9_2

Introduction

For sustainable agriculture and rural development, significant changes are required in agricultural, environmental, and macroeconomic policies at the national and international levels, in both developed and developing countries. The main goals are to boost food production and improve food security while preserving the environment to support sustainable resource management. Agriculture has a special place in all communities since it sustains rural life, despite the fact that food security remains an unfulfilled policy priority for all governments.

Continuing with conventional crop production practices is inefficient in terms of maximizing the potential yield or minimizing the production cost. All inputs, including seeds, water, fertilizer and pesticides are applied uniformly (as per recommendation) to prevent lower population stand, water deficiency, biomass growth, and yield. This is usually at a higher level to avoid any possible risk in the recommendation systems (package of practices of different crops), and farmers apply even more for a potentially high yield without any information/data regarding the input requirement. This traditional system decreases the overall input use efficiency and often results in undesirable side effects like water pollution and soil degradation. Overapplication of fertilizers results in leaching into the groundwater and polluting the same, in addition to altering the nutrient balance for better crop production (Sharma & Bali, 2018). Specifically, in Indian soils, nutrient use efficiency (NUE) is very low. It varies from 30 to 50% for nitrogen (N), 15 to 20% for phosphorus (P), 60 to 70% for potassium (K) (Sarkar et al., 2021). Therefore, this in turn has serious implications in: (a) misusing natural resources, (b) food security, and (c) achieving some sustainable development goals, especially SDG 2: Zero hunger; SDG 12: Responsible consumption and production and SDG 15: Life on land.

Lombardo et al. (2017) compared the technological revolution in agriculture, taking plowing as a method typology and working capacity of a man per hour (0.4 m × 0.2 m soil furrow slice) as a reference unit. Table 2.1 showed that there are huge differences between the productivity of a tractor compared to the productivity of a tractor with automatic drive. The difference encountered between the productivity of the tractor in 3rd and 4th is due to the increased number of plow bottoms and not to the technology. However, using an advanced technology like automatic drive, it can increase the productivity by 1.3 times.

Realizing the long-term implications of the above, researchers, industry, and farmers focused their attention on increasing the input use efficiency in agriculture. The result was the introduction of precision agriculture (PAg), about three decades back in developed countries, as a management strategy based on the crop management as per the variability and site-specific conditions (Stafford, 2000). These technologies can be implemented individually or in combination to achieve more efficiency.

The precision agriculture concept emerged in the early'90 s and was then totally focused on the study and management of spatial variability in crop production.[1] Since then, the definition has been modified or refined by many; however, ISPA definition

[1] https://www.ispag.org/about/definition

Table 2.1 Working capacity for different ploughing method typologies under various technological revolutions

Ploughing method				Working capacity, m^3/h
Man + Shovel	Volume/h	m^3/h	2.5	2
	Method efficiency		0.85	
Horse + Plow	Forward speed	km/h	3.6	230
	Method efficiency		0.80	
Tractor + Single plow bottom	Forward speed	km/h	6.0	336
	Method efficiency		0.70	
Tractor + Five plow bottom	Forward speed	km/h	6.0	1680
	Method efficiency		0.70	
Tractor + Five plow bottom + Automatic drive	Forward speed	km/h	6.0	2160
	Method efficiency		0.90	

Source Lombardo et al. (2017)

may be considered as a broad, generalized, and acceptable one.[2] "Precision agriculture is a management strategy that gathers, processes, and analyzes temporal, spatial, and individual data and combines it with other information to support management decisions according to estimated variability for improved resource use efficiency, productivity, quality, profitability, and sustainability of agricultural production [January, 2021]."

Decisions those are more effective than the ones that would be made using standard crop production practices have the potential to increase resource efficiency, lower input costs, and lessen environmental damage. They would raise crop quality and yield at the same time. Global positioning systems (GPS), geographic information systems (GIS), yield monitors, far, near, and proximal sensors, and variable-rate applicators are some of the commercial technologies currently available (Robertson et al., 2007). A study of mediterranean small farms showed that economic advantage can be achieved through the application of precision agriculture technology and processes (Loures et al., 2020). Even in sub-Saharan Africa, enhancing the precision in application of inputs and optimized management practices can significantly improve the yield potential without any significant increase in inputs (Onyango et al., 2021).

After PAg technology development, adoption/limited adoption, and realization of its importance in the sustainability of world crop production in developed countries, similar efforts have been initiated in developing countries, including Brazil, China, India, and Uruguay. Such initiatives have largely targeted specific crops which yield higher economic returns (e.g., cotton, corn, sugar cane, wheat, and rice) (Alvarez & Nuthall, 2006; Khanna et al., 1999; Maohua, 2001; Mondal et al., 2011; Silva et al.,

[2] https://www.springer.com/journal/11119/updates/17240272

2011). High cost is the most frequently mentioned obstacle, followed by the lack of skilled staff and the high costs of services. One of the main challenges for the complete adoption of these technologies is to reduce machinery and equipment acquisition costs. Apart from low-cost equipment, adopting these technologies on a large scale depends on simpler and more efficient technologies, increased perception of profitability as a result of implementing them, and training of skilled labor.

Other obstacles reported are small-scale farming and limited knowledge. Farm software adoption behavior does have similarities despite the differences in the stages of development, cultural background, and farm size had a strong influence on software use decisions. Farmers who use computers believe that they have improved their information management.

According to the threshold model, adoption occurs when some stimulus (such as profits, farm size, utility) exceeds a critical level. Advanced site-specific technologies may also be more likely to be adopted by innovative farmers with prior technical skills. Also, it occurred on larger farms with historically higher yields per acre. However, in the case of site-specific technologies, increasing returns to scale does not appear to provide large farms with a comparative advantage over small farms because of the possibility of hiring custom services for these technologies from professional applicators, crop consultants, or input supply dealers. Farmers' non-readiness in replacing existing equipment is another impediment to the adoption of PAg. However, farmers would be willing to adopt advanced technology if governments offered a cost-share subsidy of up to 50%, such as the Indian Government Model under the Sub-mission on Agriculture Mechanization (SMAM). In addition, crop prices may be increased so that farmers have enough capital with them to enable procurement, which is the least likely option in developing countries. This paper gives an overview of precision agriculture and how its adoption can help in the sustainability issues in agriculture for developing countries.

Mechanization Level in India—An Example

Farm mechanization on Indian farms began concurrently with the green revolution, although levels are still low, around 40–45%, and heavily concentrated in a small number of crops and areas. Other farm operations are at a dismally low level of mechanization compared to tillage and seed bed preparation. A representative diagram of mechanization level of different farm operations in India is shown in Fig. 2.1, and it clearly depicts the mechanization status.

The need for affordable, appropriate, precision, and energy-efficient machines and technologies for profitable production and processing of crops has been felt in recent years. This need arises due to growing worries about the availability of labor, especially during transplanting and harvesting, and the declining profitability of agriculture, focus on increased yield, decreased cultivation costs, loss prevention, and value addition through location-specific management techniques. Farmers have recognized these benefits, and agricultural mechanization is on the rise in many

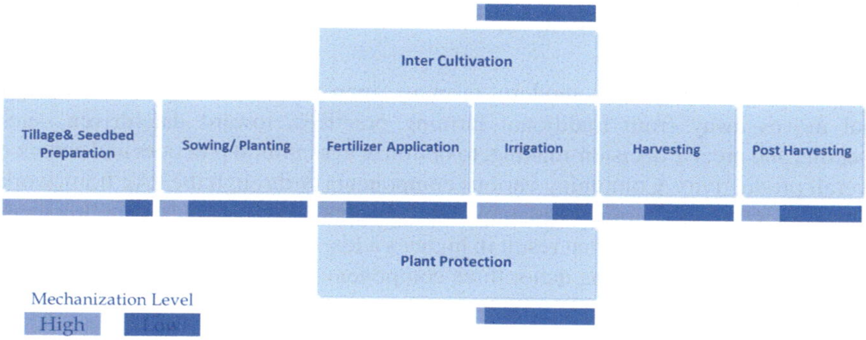

Fig. 2.1 Representative diagram of the mechanization levels of different farm operations in India

regions of the nation (Tey & Brindal, 2012). However, a multidisciplinary approach that is specific to the needs of the site should be used to consider the type of implements appropriate for cultivation and the rate at which new implements are introduced. The Indian farmers use a variety of farm tools, implements, and machinery that has been produced during the last 40 years for various farming tasks (Table 2.2). The Indian agricultural machinery market is currently expanding quickly and has significant future growth potential.

Table 2.2 Mechanization levels of farm operations in India (crop wise)

Major crop	Seedbed-preparation	Sowing/Planting/ Transplanting	Weed and pest control	Harvesting
Paddy	85–90	50–60	80–90	70–80
Wheat	90–95	80–90	70–80	80–90
Potato	90–95	80–90	80–90	70–80
Cotton	90–95	50–60	60–60	0
Gram	90–95	50–60	60–70	30–40
Millets	80–90	30–40	40–50	30–40
Sorghum	80–90	30–50	60–70	20–30
Oilseeds	80–90	30–40	60–70	20–30
Fodder crops	80–90	20–40	80–90	10–20
Horticultural crops	60–70	30–40	40–50	< 1
Vegetable crops	70–80	5–10	80–90	< 1

Source Tiwari et al. (2017)

Components of Precision Agriculture

Precision agriculture is a modern farming approach or management strategy that moves away from traditional farming practices, toward data-driven, site-specific, and precise decision-making, to optimize agricultural practices and enhance overall productivity. Combining various components or through the PAg framework (Fig. 2.2), it provides farmers with unparallel insights into their fields, enabling them to make informed choices that result in higher yields, reduced resource wastage, and improved sustainability. The major three components of PAg are:

Information

At the heart of precision agriculture lies the collection, analysis, and utilization of vast amounts of information; this includes data related to soil properties, weather patterns, crop health, and more. Advanced sensors, IoT nodes, drones, satellites, and ground-based equipment contribute to the generation of real-time data, enabling farmers to monitor and understand the conditions of their fields with exceptional accuracy. Detailed digital soil mapping provides data on soil nutrient levels, pH, moisture content, and texture, and thereby helping in making targeted decisions regarding fertilizer application for optimal growth. Also, real-time weather data, including temperature, humidity, precipitation, and wind speed, enable the farmer to predict climatic trends and potential threats like droughts or extreme weather events. Drones and satellites capture multispectral images of crops, revealing valuable insights into

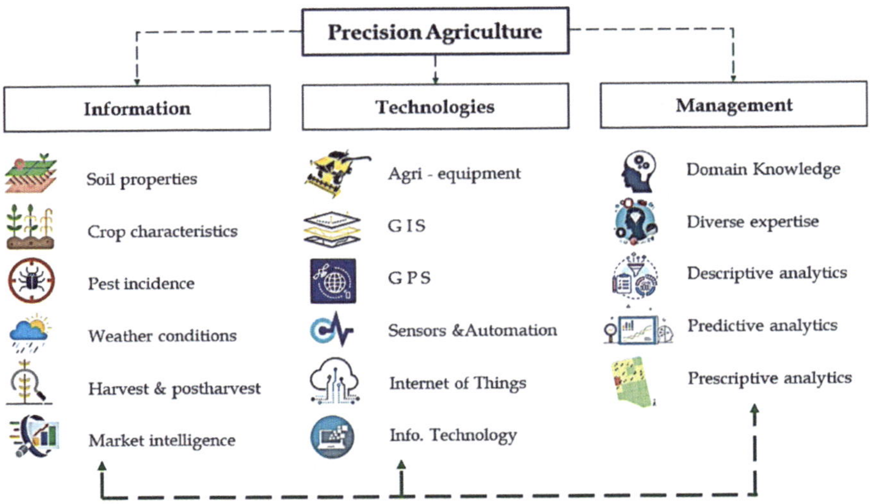

Fig. 2.2 Precision agriculture framework

their health and growth status. Vegetation indices and other metrics derived from these images indicate stress levels, pest and disease infestations, and overall crop performance.

Technology

GPS technology plays a vital role in providing accurate location information, enabling precise mapping and geo-referencing, leading to site-specific interventions and efficient resource management. Drones and satellites equipped with various sensors capture high-resolution images and data across large areas helping farmers to assess crop health, monitor plant stress, and detect pests and diseases before they cause any significant damage. The most important part of the technology is the use of automated machinery and robotic systems that can perform tasks with a high level of precision. Autonomous tractors, robotic weeders, and smart irrigation systems contribute to reducing labor demands, improving efficiency, and ensuring the accurate application of inputs. IoT devices, such as sensors and actuators, are deployed in the field as data nodes to collect real-time data on soil moisture, temperature, and other environmental variables. Using cyber-physical systems, this information is transmitted, analyzed, and visualized by agronomists, farmers, and dealers to facilitate timely decision-making.

Management

As a management strategy, precision agriculture depends on the efficient administration of the technology and data used in its application. To comprehend the enormous amounts of data created and turn them into useful insights, this necessitates using advanced algorithms and decision support systems (image processing, machine learning, artificial intelligence). PAg creates enormous amounts of data that must be efficiently arranged, saved, and processed. These datasets are handled via big data analytics and cloud computing, enabling farmers to obtain information instantly and use past data for long-term planning. Farmers can identify important patterns and trends from collected data using machine learning algorithms and data analytics. These insights enable them to make data-driven decisions like modifying irrigation schedules, optimizing fertilizer use, and detecting early signs of crop stress. Farmers can use variable rate technology (VRT) to adjust the application of inputs like fertilizers and pesticides based on the specific needs of different areas within a field (intra-field variability in both spatial and temporal dimensions) to optimize resource use and minimize waste. Farm management software is used to combine several data streams, automate procedures, and provide full farm overviews, allowing farmers to make informed decisions about crop rotation, field zoning, and overall resource management.

Challenges and Barriers in PAg Adoption

Precision agriculture has enormous potential to change farming practices and revolutionize global food production. However, its widespread adoption in developing nations confronts a number of problems and barriers. While precision agriculture has already proved its benefits in resource-rich locations, the setting of underdeveloped countries adds new complications that must be handled in order to realize its full potential. Some of the fundamental problems and barriers to precision agriculture adoption in developing nations are as follows:

High initial investment
Fragmented land holdings
Limited access to technology
Inadequate infrastructure
Access to credit
Lack of local expertise
Lack of technical knowledge
Limited internet connectivity
Limited data availability
Language and literacy barriers
Risk aversion
Inadequate government support
Cultural and social factors
Data privacy and security concerns
Climate change and environmental challenges

Recommendations for PAg Adoption in Developing Countries

Promoting and facilitating the adoption of precision agriculture in developing countries require a comprehensive and holistic approach that addresses various aspects. Implementation necessitates a multi-stakeholder approach, involving governments, international organizations, private companies, research institutions, and local communities. The following recommendations outline key strategies to overcome barriers and foster the successful implementation of precision agriculture practices in these regions:

Financial Support and Incentives

Subsidies for Technology Adoption: Governments have an important role in the introduction and adoption of innovative, sophisticated, and sustainable technology, and they should consider providing farmers with subsidies or financial incentives to purchase precision agriculture systems. These incentives can assist offset the initial capital expenditures, which are particularly high for small-scale farmers, and make the technology more accessible. Furthermore, tailored subsidies for certain technologies, such as soil and/or crop health mapping sensors or drones, could boost their acceptance and lead to wider use.

Public–Private Partnerships: Public–private partnerships can help fund research and development initiatives, provide access to cutting-edge technologies, and support training programs for farmers. These partnerships could also lead to the development of tailored financial packages or leasing options for precision agriculture equipment.

Training and Extension Services

Extension Education and Field-Level Demonstrations: Extension services can play a vital role in disseminating knowledge and best practices to farmers in remote areas. Workshops, demonstration farms, and field visits by agricultural experts can be organized to showcase the advantages of precision agriculture and its impact on productivity and sustainability.

Digital Literacy Training: As small-scale farmers may not be familiar with the application of information technology, providing basic digital literacy training will empower them to use mobile apps, access online market information, and interpret data generated by precision agriculture systems effectively.

Infrastructure Development

Improve Connectivity: Governments should prioritize expanding internet connectivity in rural areas to ensure farmers have access to online resources and real-time data. This will enable seamless communication between farmers and extension services, facilitate the exchange of information, and support the use of cloud-based data storage and analytics.

Access to Power: Access to reliable and adequate farm power is crucial for operating precision agriculture technologies. Governments and organizations can work to improve electricity infrastructure in rural regions and explore the use of renewable energy sources such as solar or wind power to support energy demand.

Research and Development

Locally Adapted Technologies: Promote research and development efforts tailored to the specific needs and challenges of developing countries. Solutions that consider local crops, soil types, and weather patterns will likely yield better results and increase the chances of successful adoption.

Climate-Resilient Technologies: Given the increased vulnerability to climate change, prioritize the development and adoption of climate-resilient precision agriculture technologies. These could include weather forecasting, early warning systems for extreme weather events, and drought-tolerant crop varieties.

Policy Support

Integration into National Agricultural Policies: Governments should integrate precision agriculture into their national agricultural policies and strategies. Including provisions that support the adoption of these technologies, setting targets for their implementation, and allocating budgetary resources will demonstrate a commitment to sustainable farming practices.

Market Access and Fair Pricing: Ensure that farmers who adopt precision agriculture technologies have fair access to markets and receive fair prices for their produce. Addressing market access challenges will create a more conducive environment for investment in precision agriculture.

Knowledge Sharing and Collaboration

Establish Knowledge Hubs: Create knowledge-sharing platforms or hubs where researchers, experts, and farmers can exchange information, experiences, and best practices related to precision agriculture. These platforms can be physical centers or digital networks that facilitate collaborative learning and problem-solving.

South–South Cooperation: Encourage South–South cooperation, where developing countries with successful precision agriculture implementations share their experiences, lessons learned, and expertise with other countries facing similar challenges. This form of collaboration can foster innovation and accelerate the adoption of sustainable farming practices.

Impact of PAg Adoption in Achieving SDGs

Precision agriculture can contribute significantly to achieving several of the United Nations' Sustainable Development Goals (SDGs) due to its inherent sustainable farming practices (Fig. 2.3). Precision agriculture's capacity to optimize resource use, increase crop yields, and improve food production efficiency can contribute

Fig. 2.3 Impact of precision agriculture adoption on SDGs

significantly to food security and hunger reduction by ensuring a stable and sustainable food supply. By precisely managing irrigation and water usage, precision agriculture not only reduces water wastage but also conserves valuable water resources, promoting sustainable water management practices (Adeyemi et al., 2017).

The integration of renewable energy sources, such as solar power, to support precision agriculture systems aligns perfectly with the goal of ensuring access to clean and sustainable energy. Moreover, precision agriculture's ability to enhance agricultural productivity and profitability fosters economic growth in the sector, creating more job opportunities and generating higher incomes. It represents a transformative technological innovation that drives agricultural modernization and fosters the development of sustainable infrastructure in rural areas. As it meets the growing demand for food, precision agriculture plays a vital role in addressing food security concerns in urban areas as well. With its resource-efficient practices that reduce waste and chemical usage, precision agriculture actively supports the promotion of sustainable consumption and production patterns. Furthermore, its climate-resilient practices and reduced greenhouse gas emissions contribute to crucial climate change mitigation and adaptation efforts in agriculture. By promoting soil health, biodiversity conservation, and sustainable land management, precision agriculture actively supports the preservation of terrestrial ecosystems and biodiversity. Lastly, precision agriculture thrives on partnerships, often involving collaborations between governments, private sector stakeholders, research institutions, and local communities, fostering synergistic efforts to achieve sustainable development goals collectively.

Precision agriculture plays a pivotal role in contributing to long-term sustainability in farming practices (Fig. 2.4). By harnessing advanced technologies, data analytics, and site-specific management, it optimizes resource utilization and minimizes environmental impacts. Through precise application of inputs such as water, fertilizers, and pesticides, wastage is significantly reduced, conserving vital resources for future generations. Additionally, precision agriculture enables farmers to implement soil and crop management practices tailored to the specific needs of their fields, promoting soil health and biodiversity. With the ability to monitor and adapt to changing climate conditions, this approach ensures resilience in the face of climate challenges, safeguarding agricultural productivity for the long run. Embracing precision agriculture is not only a path to higher yields and economic viability but also a commitment to sustainable farming practices that benefit both the environment and global food security in the years to come.

Based on precision agriculture practices, it should be possible to integrate a set of appropriate technologies to improve traditional cultivation technology to local conditions. As the present time is a century of biological and information technology revolutions, the fundamental idea for ensuring worldwide food security and human welfare will lie in scientific and technological progress. The development of precision agriculture practices will provide good opportunities for renovating conventional farming tools and technology and disseminating advanced knowledge to farmers.

The use of precisely validated zones and specific computer simulation models can increase the span between two subsequent soil samples by predicting intermediate conditions. These nutrient maps, along with easy-to-understand fertilizer recommendations for each management zone within the field, can be distributed from

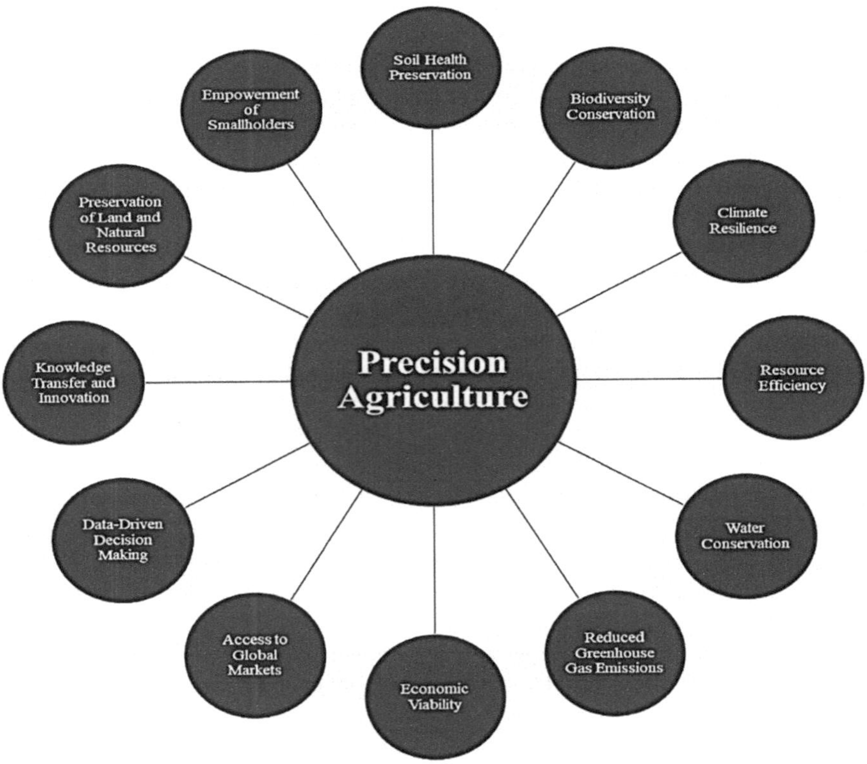

Fig. 2.4 PAg contributing to long-term sustainability

Panchayats (village regulatory bodies). PAg also demands some additional requirements, such as the development of some PAg organizations by bringing together researchers from different disciplines. Agribusiness and industry representatives, producers, and others to develop outlines as well as to carry out PAg research properly. Implementation of PAg should consider the problem of land fragmentation, lack of highly sophisticated technical centers for PAg, specific software for PAg, poor economic condition of farmers, etc. Integrating farmer knowledge, precision agriculture tools, and crop simulation modeling to evaluate management options for poor-performing areas in crop fields can be an excellent option for developed countries. Another area of improvement would be developing research tools for measuring farmers' attributes in information management preferences.

Conclusions

Adopting PAg as an enabler for sustainable agriculture in developing countries can be achieved only through increasing the level of mechanization and innovations in farm technologies, especially for small-scale farmers. Therefore, the research focus should be increasing the precision and efficiency of existing farm equipment through research and development and the free flow of technological development within the developing world. This is due to the fact that higher capacity and more advanced machines may not be the right choice for farmers who are capital deficient. Immediate economic benefits should not be a constraint for the adoption of field scale crop production system, because it will be beneficial in the long-term bringing in sustainable practices. The innovation space in each country can be enhanced by supporting and strengthening agricultural technologies (Agtech) startup especially in hardware development rather than only concentrating on the supply chain and market-based software or application (app) development. Addressing the challenges and implementing appropriate policy measures will be essential for its successful integration and contribution to sustainable development in these regions. Some of the suggestions proposed for enhancing the R&D, technology dissemination, and adoption of PAg technologies for sustainable development in developing countries are:

(a) Concentrated R&D efforts on key technologies for maximizing the input use efficiency, especially through precise application of tillage, seeds, fertilizers, pesticides, and water.
(b) Develop systems that would enable farmers to translate the extensive and multivariate spatial information into more precise judgments about input applications.
(c) Indigenous innovation in equipment design and integration of technologies.
(d) Accelerate extension activities for the use of different PAg machines, both ground and air-based.

(e) Enhanced involvement of local governments in the PAg implementation through policy directions and adequate funding to the R&D institutions and other implementing agencies.
(f) Support and strengthen the Agtech Startups.
(g) Customized farm mechanization package—crop-specific, farm holding specific, and region-specific.
(h) Paradigm shift in the curriculum for agricultural engineering and bio-systems engineering.
(i) Strengthen academia—industry–farmer interface.
(j) South–South cooperation.

Acknowledgements The author would like to thank the Council of Scientific and Industrial Research (CSIR), India for providing an opportunity to participate in the NAM Workshop.

References

Adeyemi, O., Grove, I., Peets, S., & Norton, T. (2017). Advanced monitoring and management systems for improving sustainability in precision irrigation. *Sustainability, 9*(3), 353. https://doi.org/10.3390/su9030353

Alvarez, J., & Nuthall, P. (2006). Adoption of computer-based information systems: The case of dairy farmers in Canterbury, NZ, and Florida, Uruguay. *Computers and Electronics in Agriculture, 50*(1), 48–60.

Khanna, M., Epouhe, O. E., & Hornbaker, R. (1999). Site-specific crop management: Adoption patterns and incentives. *Review of Agricultural Economics, 21*(2), 455–472.

Lombardo, S., Sarri, D., Corvo, L., Vieri, M. (2017). Approaching to the fourth agricultural revolution: Analysis of needs for the profitable introduction of smart farming in rural areas. In *Proceedings of the 8th International Conference on Information and Communication Technologies in Agriculture, Food and Environment (HAICTA2017)*, Chania, Greece, 21–24 September, 2017.

Loures, L., Chamizo, A., Ferreira, P., Loures, A., Castanho, R., & Panagopoulos, T. (2020). Assessing the effectiveness of precision agriculture management systems in Mediterranean small farms. *Sustainability, 12*(9), 3765. https://doi.org/10.3390/su12093765

Maohua, W. (2001). Possible adoption of precision agriculture for developing countries at the threshold of the new millennium. *Computers and Electronics in Agriculture, 30*, 45–50.

Mondal, P., Basu, M., Bhadoria, P. B. S., Emam, A. A., Salih, M. H., Adegbite, A. A., et al. (2011). Critical review of precision agriculture technologies and its scope of adoption in India. *American Journal of Experimental Agriculture, 1*(3), 49–68.

Onyango, C. M., Nyaga, J. M., Wetterlind, J., Söderström, M., & Piikki, K. (2021). Precision agriculture for resource use efficiency in smallholder farming systems in Sub-Saharan Africa: A systematic review. *Sustainability, 13*(3), 1158. https://doi.org/10.3390/su13031158

Robertson, M., Isbister, B., Maling, I., Oliver, Y., Wong, M., Adams, M., et al. (2007). Opportunities and constraints for managing within-field spatial variability in Western Australian grain production. *Field Crops Research, 104*(1–3), 60–67.

Sarkar, D., Rakshit, A., Al-Turki, A. I., Sayyed, R. Z., & Datta, R. (2021). Connecting bio-priming approach with integrated nutrient management for improved nutrient use efficiency in crop species. *Agriculture, 11*(4), 372. https://doi.org/10.3390/agriculture11040372

Sharma, L. K., & Bali, S. K. (2017). A review of methods to improve nitrogen use efficiency in agriculture. *Sustainability, 10*(1), 51. https://doi.org/10.3390/su10010051small

Silva, C. B., de Moraes, M. A. F. D., & Molin, J. P. (2011). Adoption and use of precision agriculture technologies in the sugarcane industry of São Paulo state, Brazil. *Precision Agriculture, 12*(1), 67–81.

Stafford, J. V. (2000). Implementing precision agriculture in the 21st century. *Journal of Agricultural Engineering Research, 76*(3), 267–275. https://doi.org/10.1006/jaer.2000.0577

Tey, Y. S., & Brindal, M. (2012). Factors influencing the adoption of precision agricultural technologies: A review for policy implications. *Precision Agriculture, 13*, 713–730. https://doi.org/10.1007/s11119-012-9273-6

Tiwari, P. S., Gurung, T. R., Sahni, R. K., & Kumar, V. (2017). Agricultural mechanization trends in SAARC region. In *Mechanization for sustainable agricultural intensification in SAARC region*. SAARC Agriculture Centre, Dhaka, 20 p.

Chapter 3
Effect of Temperature on the Treatment of Yeast Processing Effluent Using an Upflow Anaerobic Sludge Blanket (UASB) Reactor

Shepherd Manhokwe, P. Marume, R. Nyoka, and C. Zvidzai

Abstract The main objective of this study was to assess the feasibility of anaerobic treatment of yeast processing effluent in an Upflow Anaerobic Sludge Bed (UASB) reactor. Yeast-processing wastewater was obtained from a local yeast manufacturing plant and characterised. A laboratory-scale UASB reactor was used for the treatment of the effluent. Physicochemical measurements for total dissolved solids (TDS), pH, and chemical oxygen demand (COD) were made. COD was analysed using the closed reflux colorimetric method. The analysis of volatile fatty acids (VFAs) in the effluent from the UASB anaerobic digester was done using an ultra-high performance liquid chromatography (UHPLC). Trace metal ion concentration was determined using inductively coupled plasma (ICP-AES). The COD reduction varied from 32.67% at 40 °C to 19.12% at thermophilic temperature of 50°**C** during treatment of the industrial effluent. For the UASB-treated effluent, the concentration of phosphate varied from a high level of 26.14 mg/L to the lowest value of 0.841 mg/L. Biogas production peaked at 31 cm^3/day at 45 °C. The major acidogenesis products identified using UHPLC in this research were acetate and propionate. No lead was detected and zinc removal was erratic. In conclusion, temperature affected the anaerobic treatment of yeast-processing effluent. However, post-treatment of anaerobically treated effluent is recommended.

Keywords Yeast effluent · Biological treatment · UASB · Temperature

S. Manhokwe (✉) · P. Marume · R. Nyoka
Department of Food Science and Nutrition, Midlands State University, Gweru, Zimbabwe
e-mail: manhokwes@gmail.com

C. Zvidzai
Department of Biology, Bindura University of Science and Education, Bindura, Zimbabwe

X. Poshiwa et al. (eds.), *Technological Innovations for Sustainable Industrialization*,
https://doi.org/10.1007/978-981-96-7995-9_3

Introduction

The food industry contributes significantly to pollution in aquatic ecosystems (Abdulla et al., 2023; Blonskaja and Vaalu, 2009; Sun et al., 2017). The increased demand for manufactured food products coupled with industrialisation leads to the generation of large amounts of industrial effluent. However, regulatory authorities are becoming stricter on the discharge of wastewater from industries (Rajagopal et al., 2013). The application of anaerobic treatment in food processing has been documented (Austermann-Haun et al., 1999; Hampannavar & Shivayogimath, 2010; Kuo & Dow, 2017; Sigge & Britz, 2007).

The baker's yeast industry is known for the large-scale production of yeast (Igwegbe et al., 2022). The production of baker's yeast produces large amounts of wastewater from unit processes such as media preparation, fermentation, centrifugation and general cleaning in- place (CIP) (Alavijeh et al., 2017; Gengec et al., 2012; Zhou et al., 2008). The baker's yeast industry has a negative impact on the environment due to significant water consumption and high strength wastewater production. The resultant effluent contains high loads of organic matter, melanoids, and high sulphate concentrations as well as other non-biodegradable pollutants (Blonskaja et al., 2006; Liang et al., 2009). Due to the presence of melanoids, the effluent has a characteristic brownish colour (Buyukkamaci & Filibeli, 2002). Besides molasses' residuals, yeast production wastewater also contains chemicals which are added during fermentation (e.g. various salts, antifoams, propionic acid, brine, etc.), yeast metabolites, and residual yeast cells.

The UASB is a high rate reactor capable of removing organic matter in wastewater (Habeeb et al., 2010; Kumar & Bishnoi, 2017). The reactor is capable of treating high strength industrial wastewater, in addition to treating municipal wastewater (Habeeb et al., 2010; Hernández et al., 2018). The UASB consists of a tubular section, with an influent port at the bottom and an effluent port at the top of the reactor. The upper phase has a gas–liquid–solid separator to allow separation of biogas from the supernatant. The UASB has longer retention time for its biomass compared to the hydraulic retention time (HRT) of the influent (Oakley et al., 2017).UASB can be operated in psychrophilic, mesophilic, or thermophilic conditions depending on the type of matrix fed (Collivignarelli et al., 2021).

Methane is formed as a result of the synergistic relationship between microorganisms involved in hydrolysis, acidogenesis, acetogenesis, and methanogenesis (Aslanzadeh, 2014). Optimal performance of anaerobic digestion depends on several parameters (Adekunle & Okolie, 2015). The operating temperature of the reactor is one of the most important decisions that must be made in designing an anaerobic digestion system. Temperature affects the thermodynamics and kinetics of the anaerobic process. Microorganisms may work in the psychrophilic range, mesophillic range, and thermophillic range (Cheng, 2010; Drapcho et al., 2008; Kamyab & Zilouei, 2021). Therefore, there is need to optimise the digester temperature in order to maintain the ecological balance in the reactor.

Biological treatment using a bench-scale UASB was investigated for baker's yeast wastewater treatment containing highly concentrated, brown and bio-refractory compounds. The UASB reactor was subjected to various temperatures to establish optimal conditions for the removal of organic load.

Materials and Methods

Wastewater

Yeast-processing effluent was collected from a local yeast processing plant. The effluent was collected from an equalisation tank in 5 L plastic containers and characterised for physicochemical properties.

Sludge Inoculum

The inoculum for the anaerobic digestion process was obtained from an operational digester at a local brewery. The sludge was collected at the bottom of the digester in the most active zone. Sludge samples were characterised for total solids (TS) and total volatile solids (TVS). The sludge had a TS concentration of 8.2% and a TVS concentration of 2.09%.

Bioreactor

A laboratory-scale UASB reactor shown in Fig. 3.1 was used in this study. The digester had an operational volume of 1L. The reactor was jacketed to temperature adjustment during anaerobic digestion. Oxygen was prevented from entering by a U-shaped tube in the biogas collection system (Sigge & Britz, 2007). Anoxic conditions were created by sealing all openings including the lid of the reactor. Masterflex L/S peristaltic pump (Model 7554-95, Cole-Parmer, USA) was used for feeding the reactor. The wastewater was applied intermittently. The supernatant was collected from the top of the reactor and analysed to determine the treatment efficiency. During start-up, the reactors were operated at ambient temperatures with a retention time of 96 h to allow sludge adaptation to baker's yeast wastewater. When the experimental regime was completed at ambient temperature, the same procedure was carried out for reactor temperatures of 30 °C, 35 °C, 40 °C, 45 °C, and 50 °C using a jacketed reactor. The reactor was operated for 25 days for each temperature adjustment. Gas production was measured daily by using water displacement method (Abdel-Hadi, 2008; Jijaia et al., 2015).

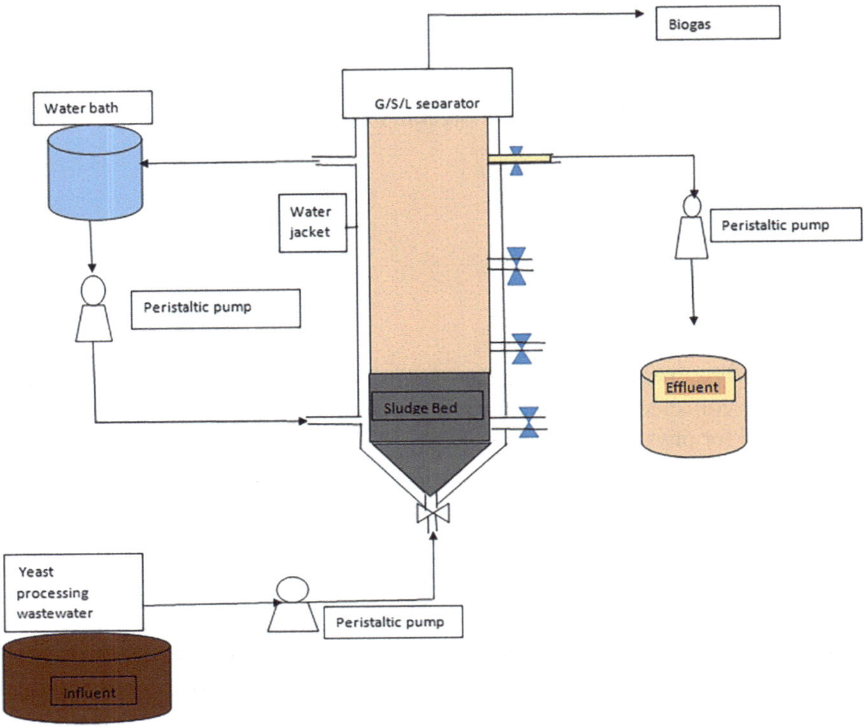

Fig. 3.1 Schematic drawing of the experimental set-up in the treatment of baker's yeast wastewater

Analyses

Physicochemical analyses for TDS, pH, sulphates, and phosphates were performed according to the Standard Methods for the Examination of Water and Wastewater (APHA, 1995) COD was analysed as per the closed reflux colorimetric method using potassium dichromate (HI93754B-25 COD MR Reagent, HANNA Instruments, USA). Samples were digested for 2 h at 150 °C in a COD reactor (Model HI 839800, HANNA Instruments, USA). COD measurement was carried out using a Multiparameter Bench photometer (Model 83214, HANNA Instruments, USA). Trace metal ion concentrations were determined using Inductively Coupled Plasma (ICP-AES) (ICAP 6000 SERIES).

Organic Acids and Metals

The analysis of Volatile Fatty Acids (VFAs) in the effluent from the UASB anaerobic digester samples at 40 °C was done using an Ultrahigh Performance Liquid Chromatography (UHPLC) system (UltiMate 3000 HPLC, Thermo Fisher Scientific). Outflow samples were collected and centrifuged in a Microcentrifuge at 4000 rpm for 5 min after which the samples were filtered through 0.45 μm pore filters. The samples were then stored at −20 °C prior to organic acid analysis. The UHPLC system equipped with the C_{18} column was calibrated using three standards of acetic acid, propionic acid and lactic acid (Sigma-Aldrich) and the corresponding calibration curve, with R^2 value 0.9988, was developed. The concentration of volatile fatty acids (VFAs) was determined employing the UHPLC system equipped with a UV–VIS detector and Hypersil GOLD column (Hypersil GOLD, 5 μm, 150 × 4.6 mm). The run time was set at 5 min. The column oven temperature was set at 30 °C, injection volume at 20 μl, and wavelength was set at 210 nm. The mobile phase A was 10 mM KH_2PO_4, pH 2.4 with phosphoric acid and mobile phase B was Acetonitrile at a flow rate of 1 ml/min. VFA concentrations were determined by comparing the peak area of calibration with that of the sample using Thermo Scientific Chromeleon Chromatography Data System (CDS) software.

Results and Discussion

Physicochemical Properties of Effluent

From Table 3.1, it is evident that the raw effluent does not meet specifications set by the Standards Association of Zimbabwe (SAZ) 588:1999, due to elevated levels of iron, zinc, manganese, COD, and phosphate which are above the permissible limits.

During the study, a decrease in COD reduction was observed at extreme temperatures as shown in Fig. 3.2. The COD reduction varied from 32.67% at 40 °C to 19.12% using a thermophilic temperature of 50 °C. The COD reduction was low compared to other studies on wastewater treatment of food-processing effluents. In a related study, the researchers achieved a maximum COD removal of 89.4% using a UASB to treat sugar industry waste (Hampannavar & Shivayogimath, 2010). Another study achieved 90% COD removal efficiency treating carbohydrate-rich effluent from the canning industry using a UASB reactor (Trnovec and Britz, 1998). After modifying operating conditions in a UASB reactor treating stillage wastewater, Mirsepasi et al. (2006) recorded an average COD removal efficiency of 55% at an OLR of 10–11 kg COD/m^3/day.

As shown in Fig. 3.3, there was no notable change in TDS. Hydrolysis, which is the first step in anaerobic digestion, transforms complex molecules into simple soluble organic molecules (Adekunle & Okolie, 2015). This contributes to the high levels of dissolved solids recorded. Researchers have shown that hydrolysis can be

Table 3.1 Yeast effluent parameters

Parameters	Mean	SAZ 588:1999 specification EMA [20:27] S.I. 6 of 2007
		Maximum permissible limit
pH @25 °C	6.74 ± 1.2	6.0–9.0
COD—mg/L	3097 ± 2510	60
Iron (as Fe^{2+}), mg L^{-1}	10.90 ± 0.92	1.0
Lead (as Pb^{2+}) mg L^{-1}	0.023 ± 0.1	0.05
Zinc (as Zn^{2+}) mg L^{-1}	10.04 ± 0.25	3.0
Copper (as Cu^{2+}) mg L^{-1}	0.26 ± 0.01	1.0
Arsenic (as As^{3+}) mg L^{-1}	ND	0.05
Cobalt (as Co^{2+}) mg L^{-1}	0.081 ± 0.01	–
Chromium (as Cr^{3+}) mg L^{-1}	0.070 ± 0.02	1.0
Manganese (as Mn^{2+}) mg L^{-1}	3.68 ± 0.14	0.1
Nickel (as Ni^{2+}) mg L^{-1}	0.27 ± 0.02	0.3
Total phosphates (as $PO4^{-3}$-) mg L^{-1}	150.06 ± 86.34	0.5
Total dissolved solids mg L^{-1}	457.5 ± 384	500
Total nitrogen—mg L^{-1}	0.612 ± 0.32	10
Chlorides (as Cl^{-}) mg L^{-1}	29.30 ± 8.94	250

ND Not detected

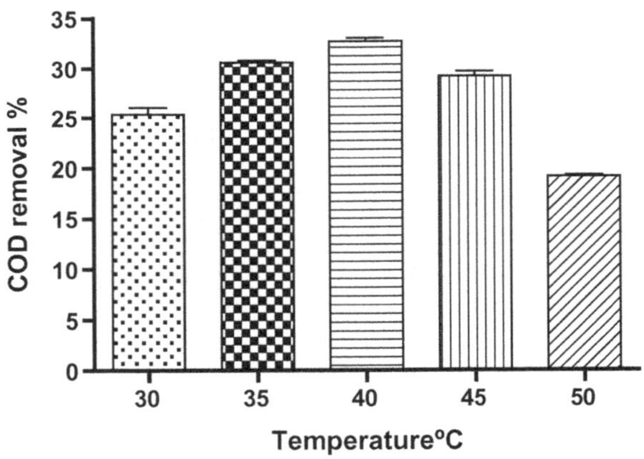

Fig. 3.2 COD reduction during treatment of yeast effluent at various temperatures in a UASB reactor

the rate-limiting step if toxic or undesirable VFAs are formed (Fernandes et al., 2009; Heo et al., 2003; Izumi et al., 2010). However, temperature had no notable effect on the levels of TDS.

The pH varied from an average of pH 4.9 at 30 °C to pH 6.9 at 40 °C, as shown in Fig. 3.4. These values at 40 °C are within the recommended statutory limits of pH 6–9 [SI 6 of 2007, Environmental Management Agency Act (20:27)]. However, most

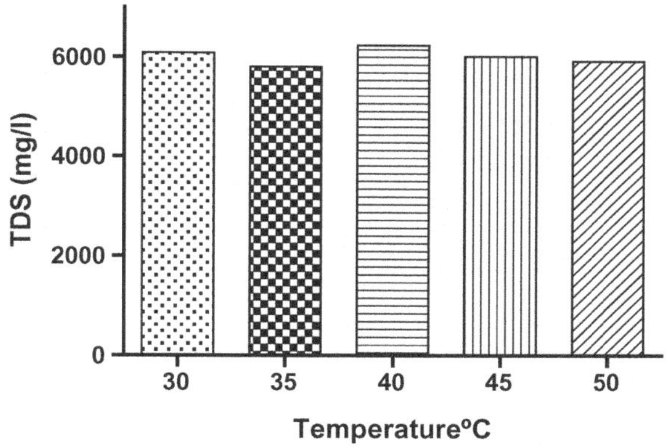

Fig. 3.3 Changes in total dissolved solids during treatment of yeast effluent at various temperatures in a UASB reactor

reactor effluent had reduced pH indicating a necessity for pH adjustment during post-treatment. The optimum pH for microbial growth should be in the range of 6.5–8.2 (Bustillo & Mahrab, 2015; Maria, 2015). A decrease in pH indicates the accumulation of VFAs from anaerobic digestion leading to reduced reactor performance and lower methane levels (Karadag et al., 2015).

The synergistic relationship between acid-formers and the methanogens should be maintained. An imbalance can trigger accumulation of VFAs which ultimately inhibit methanogens (Yu & Fang, 2003). Imbalance can occur due to excessive organic loading or changes in operating conditions which would trigger the accumulation of acidic intermediate products inhibit methanogens (Kamyab & Zilouei, 2021; Utami et al., 2016).

pH should be kept in the range of 6.8–7.4 for the optimum growth of methanogens (Sitorusa et al., 2013). At increased levels of VFAs and carbonic acid the digester acidity increases. Trimethylglcine is degraded to form carbon dioxide, methane, and ammonium (Koplima et al., 2010). The buffering effect of the ammonium formed

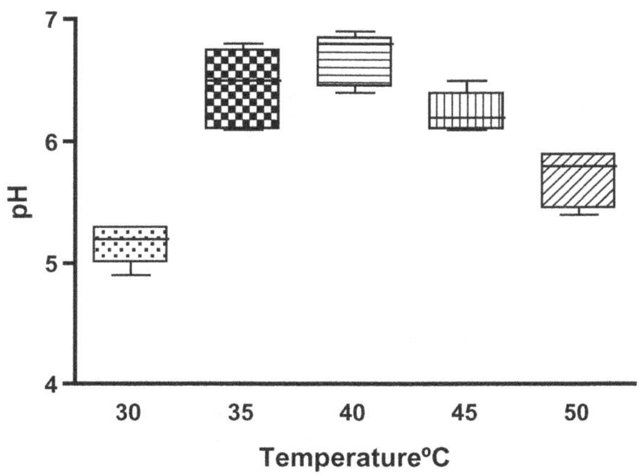

Fig. 3.4 The reactor effluent pH during treatment of yeast effluent in a UASB reactor

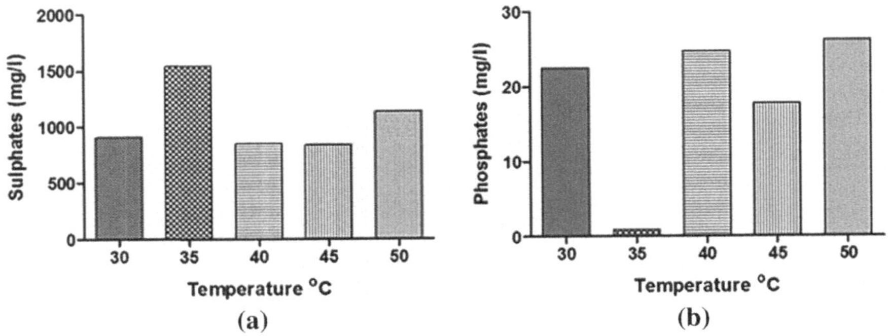

Fig. 3.5 Changes in sulphate and phosphate levels during treatment of yeast effluent at various temperatures in a UASB reactor

enables the stability of the whole system. The slight increase in pH could have resulted from the buffering effect of ammonium.

Phosphates and Sulphates

For the UASB-treated effluent the concentration of phosphate varied from a high level of 26.14 mg/L (82% reduction) to the lowest value of 0.84 mg/L (99% reduction). The lowest value attained in this study is slightly above the recommended specification of 0.5 mg/L. In a related study, the per cent removal of the phosphate concentration of the industrial effluent was found to be 53.36% (Kumar & Bishnoi, 2017). Figure 3.5 below shows the levels of sulphates (a) and phosphates (b) during anaerobic digestion of baker's yeast effluent.

For the yeast-processing effluent treatment, the sulphate concentration ranged from 909 mg/L at 30 °C to 1538 mg/L at 35 °C. The recommended limit for sulphates is 250 mg/L. Therefore, the biological process did not adequately remove the sulphates. Since methanogens and sulphate reducing bacteria compete for the same substrates, increased sulphate levels reduces the amount of methane generated (Daud et al., 2018; Farajzadehha et al., 2012; Mahmoud, 2002). This is evidenced by the decrease in biogas production to a low value of 54.8 $cm^{3/}$day at 35 °C. So, sulphate-reducing bacteria (SRB) and methanogenic bacteria compete for substrate utilization.

Biogas Production

Figure 3.6 shows the average production of biogas after the 20-day start-up period. The average volumes of biogas generated ranged from a minimum of 14 $cm^{3/}$d at

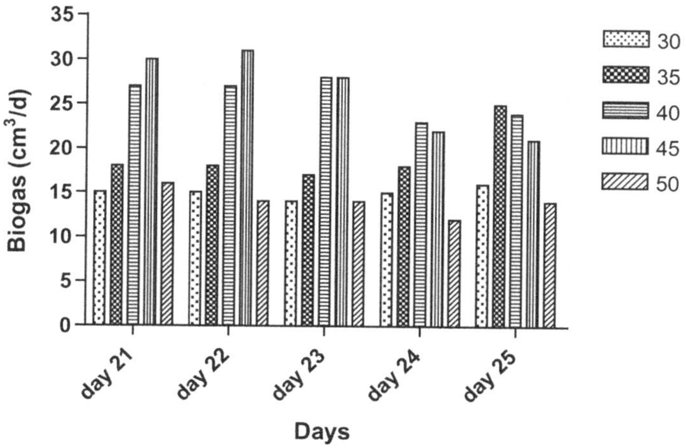

Fig. 3.6 Biogas production during steady state operation of UASB reactor for yeast effluent treatment at various temperatures

30 °C to a maximum of 31 $cm^{3/}d$ at 45 °C. Biogas production increased slightly with temperature, particularly between 40 °C and 50 °C in this study. One of the factors that enhance the ability of methanogens to produce biogas in UASB reactors is temperature (Kaviyarasan, 2014). Low temperatures have been shown to reduce the methanogen growth rates (Karadag et al., 2015). In a related study, Neena et al. (2007) reported a COD-to-biogas conversion efficiency of 81.9%. This can provide a substantial amount of energy, since anaerobic degradation stores approximately 13.5 MJ of energy per kilogram of COD (Daud et al., 2018; van Lier et al., 2015).

Organic Acids

Anaerobic digestion of organic matter leads to the formation of volatile fatty acids (VFAs), which are an indication of bacterial activity in a sample matrix as shown in Fig. 3.7. These low molecular mass carboxylic acids (C_2–C_7 monocarboxylic aliphatic acids) or short chain fatty acids (SCFAs) are important intermediates and metabolites in biological processes involving the anaerobic digestion of wastewater (Chatterjee et al., 2017; Lee et al., 2015). Acidogenesis primarily accounts for the formation of VFAs in anaerobic digesters and the VFAs are mainly acetic, propionic, butyric, and valeric acids (Chatterjee et al., 2017). The subsequent conversion of these VFAs and alcohols into methane and carbon dioxide by syntrophic acetogens and methanogenic bacteria is termed methanogenesis.

The major acidogenesis products measured using UHPLC in this research were acetate, propionate, and lactate. The average concentrations of all the VFAs recorded

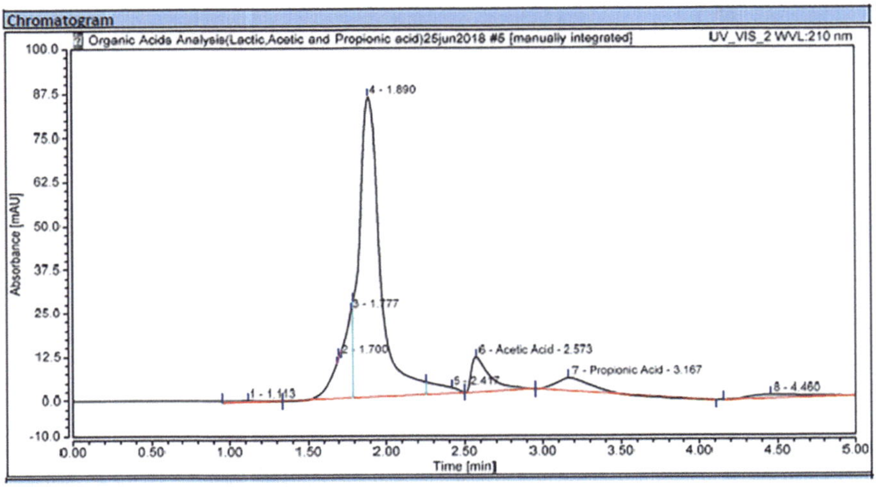

Fig. 3.7 Profiling of organic acids in a sample of raw effluent using UHPLC

in the effluent were 2743 ppm and 1326 ppm for acetic and propionic acid, respectively. During the treatment of yeast-processing effluent, the levels of acetic acid rose to 3563 ppm with a notable absence of propionic acid. No lactate was detected in both the raw and treated samples. This is attributed to the fact that VFAs are converted to acetic acid during acetogenesis, the third stage of anaerobic digestion.

Metals

For the removal of zinc, a notable result was observed at 35 °C with a concentration as low as 0.548 mg/L, as shown in Fig. 3.8. This is slightly above the recommended level of 0.5 mg/L (Zn).In a study by Atlaş (2009) heavy metals (Zn, Cr, Ni, and Cd) were shown to have a significant effect on the production of methane. No Lead (Pb) was detected in all samples analysed. However, the results indicated that UASB performance in zinc removal is independent of temperature variations.

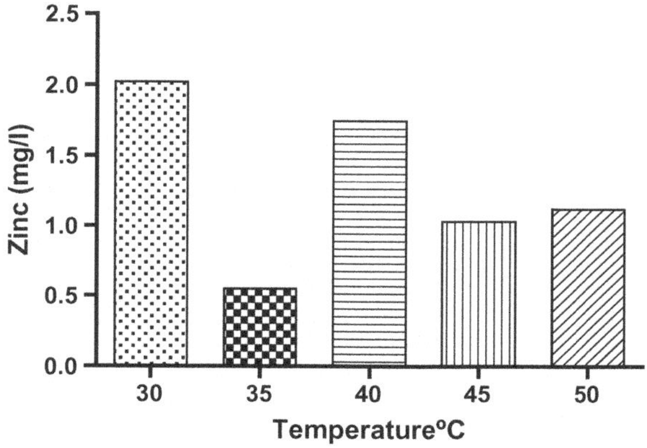

Fig. 3.8 Changes in zinc levels during treatment of yeast effluent at various temperatures in a UASB reactor

Conclusion

Temperature affects anaerobic treatment of yeast processing effluent. Relatively high COD removal rates were recorded in the thermophilic range (> 40 °C). However, reduction rates of TDS, phosphates, sulphates and zinc were erratic. Gas production was also higher in the thermophilic range with lower levels at extreme temperatures used in this study. A marked increase in acetic acid during anaerobic digestion was recorded. Therefore, anaerobic treatment of baker's yeast effluent can be done at elevated temperatures and requires post-treatment to attain stipulated values by regulatory agencies.

Acknowledgements The authors would like to thank Midlands State University for their technical support. Wastewater provided by the local yeast manufacturing plant is gratefully appreciated.

References

Abdel-Hadi, M. (2008). A simple apparatus for biogas quality determination. *MISR Journal of Agricultural Engineering, 25*(3), 1055–1066.

Abdulla, H., Zamorano, M., Rodríguez, M. L., et al. (2023). An overview of agro-food industry wastewater treatment: A bibliometric analysis and literature review. *Applied Water Science, 13*, 47. https://doi.org/10.1007/s13201-022-01857-3

Adekunle, K. F., & Okolie, J. A. (2015). A review of biochemical process of anaerobic digestion. *Advances in Bioscience and Biotechnology*. https://doi.org/10.4236/abb.2015.63020

Alavijeh, H., Sadeghi, M., Rajaeieh, M., Moheb, A., & Sadani, M. (2017). Integrated ultrafiltration membranes and chemical coagulation for treatment of baker's yeast wastewater. *Journal of Membrane Science and Technology*. https://doi.org/10.4172/2155-9589.1000173

Altaş, L. (2009). Inhibitory effect of heavy metals on methane-producing anaerobic granular sludge. *Journal of Hazardous Materials, 162*(2–3), 1551–1556.

APHA. (1995).*Standard methods for the examination of water and wastewater*. American Public Health Association/American Water Works Association/Water Environment Federation.

Aslanzadeh S., Berg A., Taherzadeh M.J., Horváth S.I. (2014). Biogas production from N-methylmorpholine-N-oxide (NMMO) pretreated forest residues. *Applied Biochemistry and Biotechnolology, 172*(6):2998–3008. https://doi.org/10.1007/s12010-014-0747-z

Austermann-Haun, U., Meyer, H., Seyfried, C. F., & Ronwinkel, H. (1999). Full scale experiences with anaerobic/aerobic treatment plants in the food and beverage industry. *Water Science and Technology, 40*(1), 305–312.

Blonskaja, V., Kamenev, I., & Zub, S. (2006). Possibilities of using ozone for the treatment of wastewater from the yeast industry. *Proceedings of the Estonian Academy of Sciences, 55*(1), 29–39.

Blonskaja, V., & Vaalu, T. (2009). Possible ways for post-treatment of biologically treated wastewater from yeast factory. *Journal of Environmental Engineering and Landscape Management, 17*(4), 189–197. https://doi.org/10.3846/1648-6897.2009.17.189-197

Bustillo, L., & Mahrab, M. (2015). Slaughterhouse wastewater characteristics, treatment, and management in the meat processing industry: A review on trends and advances. *Journal of Environmental Management, 161*, 287–302.

Buyukkamaci, N., & Filibeli, A. (2002). Concentrated wastewater treatment studies using an anaerobic hybrid reactor. *Process Biochemistry, 38*, 771–775.

Chatterjee, B., Radhakrishnan, L., & Mazumde, D. (2017). New approach for determination of volatile fatty acid in anaerobic digester sample. *Environmental Engineering Science*. https://doi.org/10.1089/ees.2017.0190

Cheng, J. (2010). Anaerobic digestion for biogas production. In J. Cheng (Ed.), *Biomass to renewable energy processes* (pp. 151–208). CRC Press.

Collivignarelli, M. C., Abbà, A., Caccamo, F. M., Calatroni, S., Torretta, V., Katsoyiannis, I. A., CarnevaleMiino, M., & Rada, E. C. (2021). Applications of up-flow anaerobic sludge blanket (UASB) and characteristics of its microbial community: A review of bibliometric trend and recent findings. *International Journal of Environmental Research Public Health, 18*, 10326. https://doi.org/10.3390/ijerph181910326

Daud, M. K., Rizvi, H., Akramm, F., Ali, S., Rizwan, M., Nafees, M., & Shuijin, Z. (2018). Review of upflow anaerobic sludge blanket reactor technology: Effect of different parameters and developments for domestic wastewater treatment. *Hindawi Journal of Chemistry, 1596319*, 1–13. https://doi.org/10.1155/2018/1596319

Drapcho, C. M., Nhuan, N. P., & Walker, T. H. (2008). *Biofuels Engineering Process Technology*. McGraw-Hill Companies Inc.

Environmental Management Act [Chapter 20:27].

Farajzadehha, S., Mirbagheria, S., Farajzadehha, S., & Shayegan, J. (2012). Lab scale study of HRT and OLR optimization in UASB reactor for pretreating fortified wastewater in various operational temperatures. *APCBEE Procedia, 1*, 90–95.

Fernandes, T. V., KlaasseBos, G. J., Zeeman, G., Sander, J. P. M., & Lier, J. B. (2009). Effects of thermo-chemical pretreatment on anaerobic biodegradability and hydrolysis of lingocellulosic biomass. *Bioresource Technology, 100*, 2575–2579. https://doi.org/10.1016/j.biortech.2008.12.012

Gengec, E., Kobya, M., Demirbas, E., Akyol, A., & Oktor, K. (2012). Optimization of baker's yeast wastewater using response surface methodology by electrocoagulation. *Desalination, 286*, 200–209.

Habeeb, S. A., Latiff, A. B. A., & Ahmad, Z. B. (2010). A review on properties of the digestion process in the up-flow anaerobic sludge bed (UASB) reactor. *Canadian Journal on Environmental, Construction and Civil Engineering, 1*(3), 58–70.

Hampannavar, U. S., & Shivayogimath, C. B. (2010). Anaerobic treatment of sugar industry wastewater by Upflow anaerobic sludge blanket reactor at ambient temperature. *International Journal of Environmental Sciences, 1*(4), 631–639.

Heo, N. H., Park, S. C., Lee, J. S., & Kang, H. (2003). Solubilization of waste activated sludge by alkaline pretreatment and biochemical methane (BMP) tests for anaerobic co-digestion of municipal organic waste. *Water Science and Technology, 48*, 211–219.

Hernández, D. M., Hurtado, A. P., & Chaparro, T. R. (2018). Hydrogen production in a novel configuration of UASB reactor under different hydraulic retention time. *Dyna, 85*(205), 157–162.

Igwegbe, C., Ifeoma, O., Iwuozor, K., Ghosh, S., Kurniawan, S., Rangabhashiyam, S., Kanaoujiya, R., & Ighalo, O. J. (2022). Treatment technologies for bakers' yeast production wastewater. *Environmental Science and Pollution Research*. https://doi.org/10.1007/s11356-021-17992-4

Izumi, K., Okishio, Y. K., Niwa, C., Yamamoto, S., & Toda, T. (2010). Effects of particle size on anaerobic digestion of food waste. *International Biodeterioration & Biodegradation, 64*, 601–608.

Jijaia, S., Srisuwana, G., O-Thong, S., Ismaile, N., & Siripatan, C. (2015). Effect of granule sizes on the performance of upflow anaerobic sludge blanket (UASB) reactors for cassava wastewater treatment. In *International Conference on Alternative Energy in Developing Countries and Emerging Economies Energy. Procedia, 79*, 90–97.

Kamyab, B., & Zilouei, H. (2021). Investigating the efficiency of biogas production using modelling anaerobic digestion of baker's yeast wastewater on two-stage mixed-UASB reactor. *Fuel, 285*, 119198. https://doi.org/10.1016/j.fuel.2020.119198

Karadag, D., Koroglu, O. E., Ozkaya, B., Cakmakci, M., Heaven, S., Banks, C., & Serna-Maza, A. (2015). Anaerobic granular reactors for the treatment of dairy wastewater: A review. *International Journal of Dairy Technology, 68*, 1–12. https://doi.org/10.1111/1471-0307.12252

Kaviyarasan, K. (2014). Application of UASB reactor in industrial wastewater treatment—A review. *International Journal of Scientific & Engineering Research, 5*(1), 584–589.

Koplima, M., Menert, A., Blonskaja, V., Kurissoo, T., Zub, S., Saareleht, M., Vaarmets, E., & Menert, G. T. (2010). Liquid and gas chromatographic studies of the anaerobic degradation of baker's yeast wastewater. *Procedia Chemistry, 2*, 120–129.

Kumar, S. S., & Bishnoi, N. R. (2017). Treatment of food industry effluent using upflow anaerobic sludge blanket reactor. *Biochemical and Cellular Archives, 17*(1), 73–77.

Kuo, J., & Dow, J. (2017). Biogas production from anaerobic digestion of food waste and relevant air quality implications. *Journal of the Air and Waste Management Association, 67*(9), 1000–1011. https://doi.org/10.1080/10962247.2017.1316326

Lee, D., Lee, S., & Bae, J. (2015). Effect of volatile fatty acid concentration on anaerobic degradation rate from field anaerobic digestion facilities treating food waste leachate in South Korea. *Journal of Chemistry*. https://doi.org/10.1155/2015/640717

Liang, Z., Wang, Y., Zhou, Y., & Liu, H. (2009). Coagulation removal of melanoidins from biologically treated molasses wastewater using ferric chloride. *Chemical Engineering Journal, 152*, 88–94.

Mahmoud N. (2002). *Anaerobic pre-treatment of sewage under low temperature (15 °C) conditions in an integrated UASB-Digester system*. Thesis, Department of Environmental Technology and the Department of Agro-Technology and Food Sciences, Wageningen University, Wageningen, The Netherlands

Maria, J. S. (2015). TratamentoAnaeróbio do Efluente da Indústria de Lacticíniosusando um Reactor UASB. Dissertação de Mestradoem Engenharia do Ambiente, Universidade Agostinho Neto, Luanda, Angola.

Mirsepasi, A., Honary, H. R., Mesdaghinia, A. R., Mahvi, A. H., Vahid, H., & Karyab, H. (2006). Performance evaluation of full scale UASB reactor in treating stillage. *Wastewater, Iranian Journal of Environmental Health, Science and Engineering, 3*(2), 79–84.

Neena, C., Ambily, A. P., & Jisha, M. S. (2007). Anaerobic degradation of coconut husk leachate using UASB-reactor. *Journal of Environmental Biology, 28*(3), 611–615.

Oakley, S., Von Sperling, M., & Verbyla, M. (2017). Anaerobic sludge blanket reactors. In J. B. Rose & B. Jiménez-Cisneros (Eds.) *Global Water Pathogen Project*. http://www.waterpathogens.org

Rajagopal, R., Saady, N. M. C., Torrijos, M., Thanikal, J. V., & Hung, Y. (2013). Sustainable agro-food industrial wastewater treatment using high rate anaerobic process. *Water, 5*, 292–311. https://doi.org/10.3390/w5010292

SAZ. (1999). Standard Association of Zimbabwe: ZWS 588:1999 Zimbabwean Standard Specification for Effluent Water.
Sigge, G. O., & Britz, T. J. (2007). UASB treatment of a highly alkaline fruit-cannery lye-peeling wastewater. *Water SA, 33*(2), 275–278.
Sitorusa, B., Seno, S., & Panjaitanc, D. (2013). Biogas recovery from anaerobic digestion process of mixed fruit–vegetable wastes. In *International Conference on Sustainable Energy Engineering and Application, Energy Procedia, 32*, 176–182.
Sun, F., Yun, D., & Yu, X. (2017). Air pollution, food production and food security: A review from the perspective of food system. *Journal of Integrative Agriculture, 16*(12), 60345–60347.
Trnovec W., & Britz T.J. (1998). Influence of organic loading rate and hydraulic retention time on the efficiency of a UASB bioreactor treating a canning factory effluent. *Water SA, 24*,(2), ISSN 3784738. http://hdl.handle.net/10019.1/10836
Utami, I., Redjeki, S., Astuti, D. H., & San, I. (2016). Biogas production and removal COD–BOD and TSS from wastewater industrial alcohol (vinasse) by modified UASB bioreactor. *MATEC Web of Conferences, 58*, 01005.
van Lier, J. B., Van Der Zee, F. P., Frijters, C. T., & Ersahin, M. E. (2015). Celebrating 40 years anaerobic sludge bed reactors for industrial wastewater treatment. *Reviews in Environmental Science and Bio-Technology*. https://doi.org/10.1007/s11157-015-9375-5
Yasar, A., & Tabinda, A. B. (2010). Anaerobic treatment of industrial wastewater by UASB reactor integrated with chemical oxidation processes; an overview. *Polish Journal of Environmental Studies, 19*(5), 1051–1061.
Yu H.Q., & Fang H.H.P. (2003). Acidogenesis of gelatin-rich wastewater in an upflow anaerobic reactor: influence of pH and temperature. *Water Research, 37*, 55–66.
Zhou, Y., Liang, Z., & Wang, Y. (2008). Decolorization and COD removal of secondary yeast wastewater effluents by coagulation using aluminum sulphate. *Desalination, 225*, 301–311.

Part II
Research, Innovation and Entrepreneurship

Chapter 4
Autonomous Motion Planning for 5250 LabVolt Robot

Maithem Sabri Jaber Al-Zubaidy, Hussein Tbena Kadhim, and Omar Qahtan Ahmed

Abstract The usage of robots in businesses and in important technology applications has been observed to have a significant impact in the modern world. Humans greatly benefit from the usage of autonomous systems and from their effective performance in hazardous tasks. For an autonomous robot, obstacle detection is regarded as a very important criterion. An autonomous robot that can recognize the existence of an object or obstacle and avoid colliding with it is called as an obstacle avoidance robot. This keeps both the robot and the environment safe. The right sensors and robot design must be developed for an obstacle avoidance robot. This study focuses on developing a robot system (LabVolt5250) to identify obstacles and steer clear of collisions. A microcontroller named Arduino Uno, an ultrasonic sensor called HC-SR04, and robot movement are controlled by an Arduino Uno microcontroller, a digital bus, and an uncharted environment. The robot's two attached ultrasonic sensors detect the existence of obstacles and transmit the detected information to the microcontroller. The robot will move forward, backward, right, and left depending on the data the robot controller receives. The key functions of an Arduino microcontroller are distance calculation and achieving the necessary detection and avoidance operation.

Keywords Ultrasonic sensor · Robot · Distance · Autonomous systems

M. S. J. Al-Zubaidy (✉) · O. Q. Ahmed
Information and Communication Technology Research Center, Scientific Research Commission, Ministry of Higher Education and Scientific Research, Baghdad, Iraq
e-mail: maithemsabri71@gmail.com

H. T. Kadhim
Space Research and Technology Center, Scientific Research Commission, Ministry of Higher Education and Scientific Research, Baghdad, Iraq

X. Poshiwa et al. (eds.), *Technological Innovations for Sustainable Industrialization*,
https://doi.org/10.1007/978-981-96-7995-9_4

Introduction

Modern manufacturing production lines regularly employ industrial robots, particularly for tasks requiring high precision or those that must be completed in risky or dynamic circumstances. Welding, pick-and-place work, and spray painting are common processes in manufacturing settings where industrial robots are frequently used. Since the environment is always changing, industrial robots cannot perform the majority of these tasks without the use of sufficient sensors. The distance sensor is one of the main sensor type used in dynamic work environments. The primary driver behind the quick growth of distance sensor applications is the accessibility of fast processors and inexpensive, power-efficient sensors (Gowtham, 2016; Khine et al., 2019). Range sensors are also capable of gathering a substantial amount of environmental data from a field. A distance sensor is therefore, frequently employed to deliver measurement data and to improve the robot's interaction with its surroundings. The following three components make up the typical distance sensing system (Gowtham, 2016):

- Choosing a distance sensor is part of the field acquisition process.
- Data bus devices, including TTL, USB, parallel, and serial connections, make it possible to connect the robot controller and microcontroller through an interface. This paper describes how to integrate a remote system with the laboratory robot. The robot is designed to approach, grasp, and locate a target.
- Tools for remote processing: These include processing units and computer programming languages like C++, Python.

The developed integration technique is described in the following sections. The hardware architecture of the system is described first. An explanation of the sensor calibration process is then provided. The software concept is then outlined, and the presentation of the experimental findings is presented (David et al., 2016; Kolapo et al., 2018; Sanjyoti & Isha Malhotra, 2020).

System Hardware Architecture

The hardware architecture of the system is depicted in Fig. 4.1 and consists of a LabVoltrobot model 5250 system and a distance measurement system made up of an ultrasonic sensor and a distance-processing unit.

a. **Robot System Integration**

The LabVolt robot used in this work, Model 5250, is shown in Fig. 4.2. It has five rotational axes and is powered by servo actuators: the base, shoulder, elbow, wrist, and gripper. The motor that rotates the shoulder is housed in the base, which also supports the arm mechanism. The motors that move the other parts of the arm are housed in the shoulder (Ahmed et al., 2011).

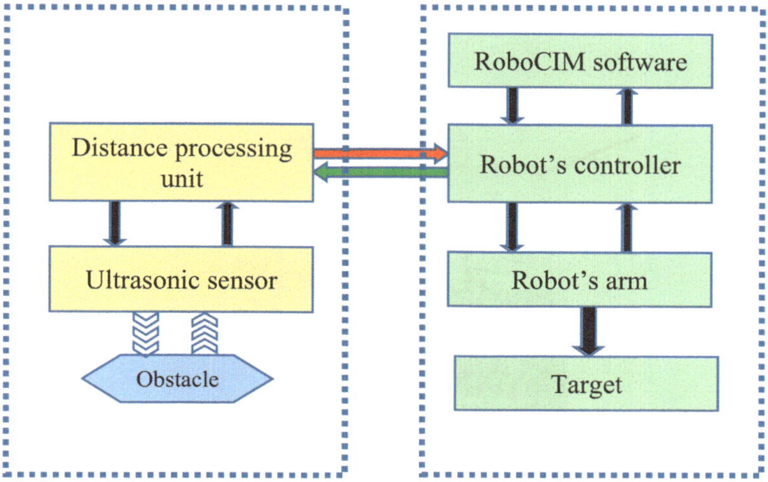

Fig. 4.1 System hardware architecture

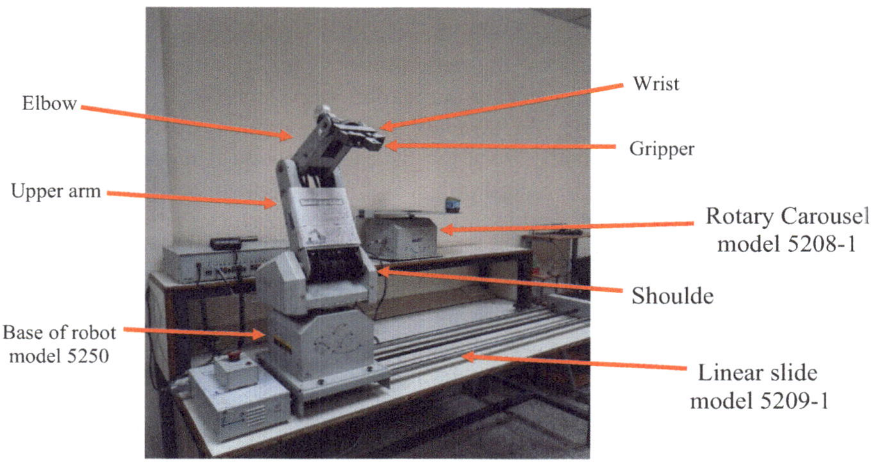

Fig. 4.2 Robot system integration

The forearm, wrist, and gripper movement mechanism are housed in the bottom portion of the upper arm. The forearm can move up and down relative to the upper arm. The gripper's position and direction are controlled by the wrist, which can rotate 360 degrees in both clockwise and counterclockwise directions. The robot's arm is equipped with a two-fingered parallel jaw gripper that allows it to pick up and move objects like tools, parts, and materials (Sanjyoti & Isha, 2020). Figure 4.3 illustrates how the controller can be integrated with external subsystems using its 16 TTL input and output channels (Ahmed et al., 2011; Mohammed et al., 2017).

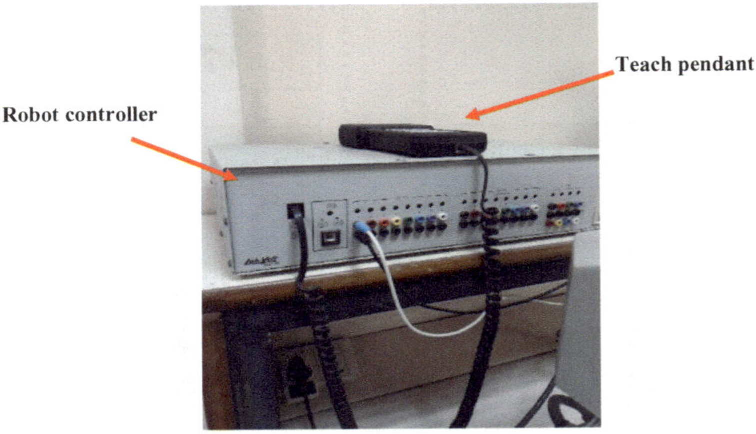

Fig. 4.3 The teach pendant and robot controller system

The RoboCIM simulation software or a handheld terminal (teach pendant) can be used to program this robot. The robot's articulation is managed using the teach pendant. Recording points and utilizing them to alter a robot program that governs the robot's movements allows for the definition of object positions. All regulated orders that assist the robot in carrying out its duties are included in the RoboCIM simulation program. It enables the development of a program, execution of a test in simulation mode, and finally execution of a test in control mode using a real robot. The software is used to record points and then use them to alter a program, much like the teach pendant. In this instance, the robot performs routine tasks. As a result, unless the recorded points are adjusted appropriately, the robot will not be able to grip the object if its placement within the robot workspace changes. The robot's software workspace is depicted in Fig. 4.4 (David et al., 2016). In automated manufacturing systems, many kinds of linear slides are employed for the automatic transfer of robots and parts along predetermined pathways. In this study, the robot is moved using a LabVoltLinear Slide Model 5209-1, and the target is rotated to the proper position using a Rotary Carousel Model 5208-1. TTL input and output ports are used to integrate these accessories with the robot controller (Ahmed et al., 2011; Mohammed et al., 2017; Mohammad, 2015; Vijayalaxmi et al., 2013).

b. **Distance System Configuration**

As previously noted, sensors, range processing units, and data bus tools are the three main components of a distance sensing system. As the range sensor for this project, an Ultrasonic Sensor HC-SR04 has been utilized. This affordable sensor, as illustrated in Fig. 4.5a, offers non-contact measuring functionality from 2 to 400 cm with a range accuracy of up to 3 mm. It is connected to the microcontroller through a high-speed port. An Arduino Uno is the microcontroller used in this project. Arduino uses Open-source technology. The Arduino boards are simple to program using the Arduino

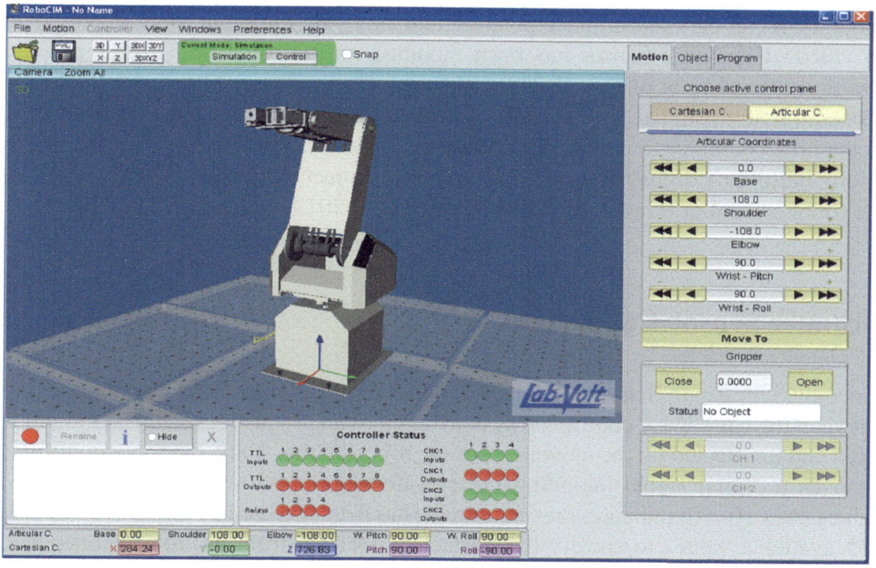

Fig. 4.4 The RoboCIM main window

Software (IDE). The Arduino Uno is based on the ATmega328P microcontroller, which has 32 kB of flash memory. It includes 14 digital input/output pins, and as Fig. 4.5b shows, six of them can be utilized as pulse width modulation outputs. The board can be powered and programmed by connecting it to a computer using a Universal Serial Bus (USB) cable; the suggested input voltage range is 7–12 V. Less than 7 V may cause it to become unstable, and over 12 V may damage the board by overheating the voltage regulator.

There are two ways to place the sensor in relation to the robot in a robotic environment. The sensor can be positioned on the robot arm so that it moves with the

(a) **(b)**

Fig. 4.5 **a** Distance sensor, **b** Arduino UNO board

robot, or it can be fixed in place position and stay still while the robot moves. The HC-SR04 Ultrasonic Sensor has been mounted on the robot arm in this project, giving it a good view of the robot's workspace. C++ software libraries for distance processing are installed on the microcontroller. A data bus tool must be connected to the host computer to enable communication with other subsystems. Eight of the microcontroller's digital ports were used in this project to transmit digital signals to the robot controller (Gowtham, 2016; Khine et al., 2019; Kolapo et al., 2018).

Distance Sensor Calibration

Since the Ultrasonic Sensor HC-SR04 is attached to the robot arm, the relationship between the workspace coordinate system and the range sensor coordinate system is fixed and unaffected by the movement of the robot. In this study, the robot controller receives a digital signal including binary information, allowing for 256 possible values. Zero ($0 = 0$) indicates that no obstacle has been detected by the sensor, leaving 255 available values to specify the obstacle's distance from the robot workspace. Ultrasonic sensor tests were conducted to determine how far from an obstacle the robot should stop. Tests were conducted to determine the sensor's accuracy. A ruler was used to measure the distance from the sensor to an obstacle, and the measured value was then compared with the distance reported by the sensor. The results of this calibration are displayed in Fig. 4.6.

The outcome demonstrated that the sensors are accurate except when an object is positioned very close to the sensor. The sensors' detection range, according to the manufacturer's specifications, is between 2 and 400 cm. Table 4.1 displays the outcome.

Fig. 4.6 Distance calibration of HC-SR04 sensor

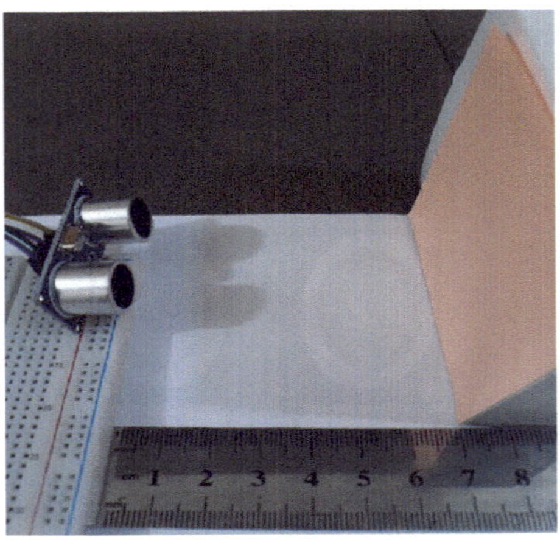

Table 4.1 The result

Distance measured by ruler (cm)	Distance measured by sensor (cm)
2	2
5	5
7	7
25	25.1
60	60.1

Software Development

Figure 4.7 shows a flowchart that illustrates the process of robot control based on the range sensing system. Distance processing is carried out in this study. It should be emphasized, however, that RoboCIM is used to program the robot (Zahraa et al., 2015).

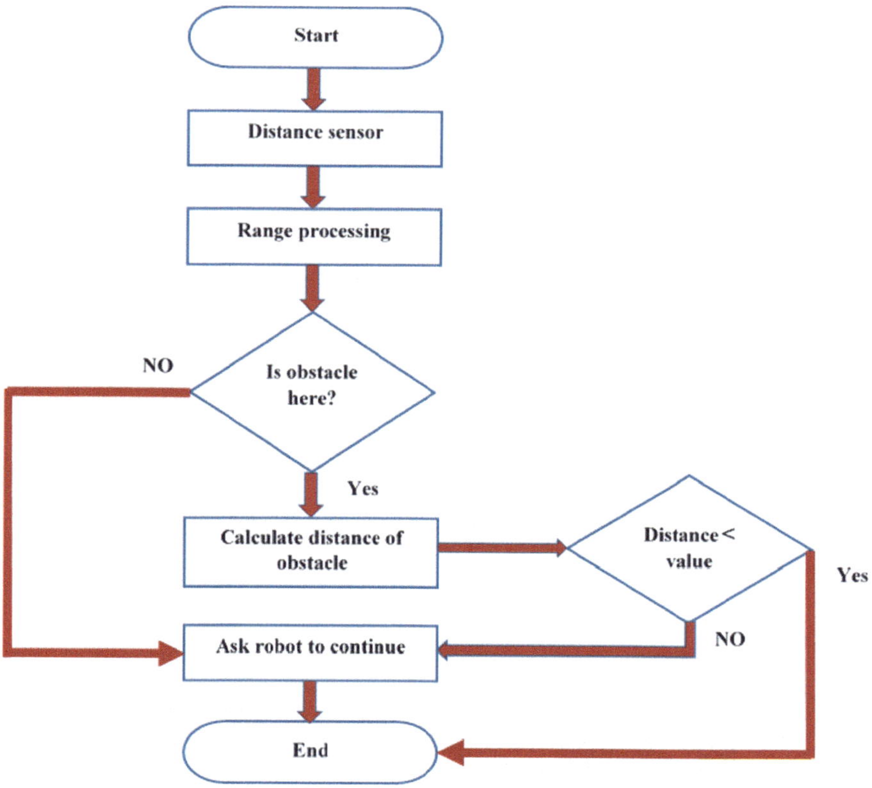

Fig. 4.7 Flowchart of the RoboCIM and distance program

To explain the algorithm that the robot employs, a flowchart was created. The robot begins by moving forward. The robot will stop, sound an alarm, and then readjust its heading if it detects an obstruction in its path. If no obstruction is detected, it keeps moving forward. When there are obstructions on both sides, the robot inspects them, and if there is a clear path, it turns and scans its right and left sides again. The robot would have to be manually repositioned if there are impediments all around it, because the motors would stop in that case.

a. **Distance Processing and Obstacles Detections**

The robot (LabVolt5250) was faced with obstacles of varied sizes and shapes to determine which objects the sensors could detect. Figure 4.8 illustrates how the sensors were attached to the robot's arm on both sides. This configuration made it possible to investigate the sensors' capacity to identify small obstructions in their path. To begin this test, an obstruction was placed in front of the sensor. The robot's range sensor had no trouble detecting the obstruction. In front of the building, another obstruction was positioned at a different angle and was once again quickly found. The results demonstrated that the sensors could detect the obstacles. The results of this test were helpful in determining which obstacles to use in the subsequently-built obstacle course and what changes could be made to the robot (Kolapo et al., 2018; Zahraa et al., 2015).

b. **Communication Program**

The obstacle distance computed in the previous phase was utilized to develop software that would generate and send digital signals to the robot controller through

Fig. 4.8 The placement of range sensors on both sides of the robot's arm

the TTL interface. The distance to the obstacle is represented by the value of the digital signal. A total of 256 values exist, corresponding to 255 distance values and the value 0 for no obstacle (Table 4.2). When there is no obstacle, the digital output from the distance system will be 0, while at the maximum distance of 400 (mm), the digital output will be 255, or FF in hexadecimal, which is the maximum value for one byte in digital form that will be transmitted to the eight TTL input channels of the controller for robot 5250. In order to match these values, a transform equation (4.1) was utilized, as shown below, and the transformation has a 2 (cm) resolution (Ahmed et al., 2011; Mohammed et al., 2017; Vijayalaxmi et al., 2013).

$$\text{Digital output distance} = \left|\text{Real input distance} \times \frac{255}{400}\right|, \tag{4.1}$$

where:

Digital output distance: value of the distance saved in one byte.
Real input distance: distance of the obstacle from the ultrasonic sensor (cm).

c. **Robot Program**

Robotics software, or "RoboCIM," enables the robot to carry out repetitive tasks. When a robot is outfitted with a range sensing system, the controller can adjust the robot's operations to suit the demands of the work environment. As previously stated, eight digital signals are delivered through TTL input channels to the robot controller. As a result, the robot software can accommodate 256 different distances in the environment. Because RoboCIM simulation software may interact with its environment via TTL input and output channels, a variety of impediments will be introduced to the software by using these TTL input channels. As a result, the robot's operation would be safer (The StaffofLabtd, 2006; Zahraa et al., 2015).

Table 4.2 Obstacle location and corresponding digital signal values

Real input distance (cm)	Digital output distance (cm)	Signal values in (binary number)	Signal values in (hexadecimal number)
0	0	0000 0000	00
1	0	0000 0000	01
2	1	0000 0001	01
3	1	0000 0001	01
4	2	0000 0010	02
5	2	0000 0010	02
	|	|	
200	127	1111 1110	7F
300	191	1011 1111	BF
400	255	1111 1111	FF

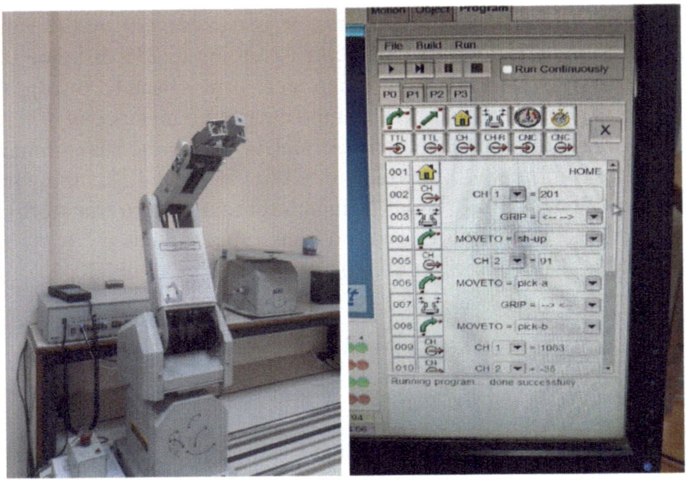

Fig. 4.9 The LabVolt robot model 5250 during an experiment test

Experimental Work

The method was tested using two obstacles placed at various distances. They were 5 cm away and 100 cm away, respectively. These obstacles were placed at random on two of the numerous predetermined spots chosen by the robot. The procedure began with readings from the sensor being sent to the Arduino UNO microcontroller for range processing. The obstacle's location in the workspace was identified as a result of the distance processing operation, being conducted, which recognized and computed the obstacle at the desired distance in the field. The corresponding digital signals, in this case as a TTL signal, were then transmitted to the robot controller. As can be seen in Fig. 4.9, the robot then approached, gripped, and placed the object. The procedure was then carried out again or stopped. The fact that every step was carried out successfully demonstrates the effectiveness of the methods used.

Conclusion

In order to validate the proposed method for integrating the simulation software with the range sensing system, experiments were conducted. Additionally, it was discovered that placing an obstacle close to the robot (5 cm) produces the best results. A range of tasks, such as pick-and-place operations, packing, processing, and quality control, could be performed using the system that was created. As an extension of this project, it is planned to improve the range system to handle moving obstacles in the workspace. In this case, it is important to consider the speed of the robot's arm relative to the speed of the obstacle.

References

Ahmed, Sh. K., Ganesh, K., & Majid, T. (2011). An approach for integration of industrial robot with vision system and simulation software. *International Journal of Information and Communication Engineering, World Academy of Science, Engineering and Technology, 5*(10).

David, G., Joshué, P., Vicente, M., & Fawzi, N. (2016). A review of motion planning techniques for automated vehicles. *IEEE Transactions on Intelligent Transportation Systems, 17*(4).

Gowtham, R. (2016). Obstacle detection and avoidance methods for autonomous mobile robot. *International Journal of Science and Research, 5*, 1245–1249.

Khine, M. M., Thae E. A., Hnin, & N. Y. P. (2019). Automatic obstacle avoidance robot vehicle. *IJCIRAS1342, 2*, 112–115.

Kolapo, S. A., Moses, O. O., Akinola, S. O., Michael, A. O., Anthony, C. M., Oladimeji, A., & Samuel, O. D. (2018). Development of an Arduino-based obstacle avoidance robotics system for an unarmed vehicle. *APRN journal of Engineering and Applied Sciences, 13*(3), 886–892.

Mohammad, N. (2015). Development of an obstacle avoiding robot (Vol. 5, No. 2). Microcontroller Lab, Univ. Pembangunan Jaya, Tangerang Selatan, Indonesia.

Mohammed, E., Nabil, D., & Mahmoud, O. (2017). Open source interfacing for LabVolt 5250 robotic arm. In *1st Conference of Industrial Technology*.

Sanjyoti, J., & Isha, M. (2020). A review on obstacle avoidance techniques on self-driving vehicle. *International Journal of Advanced Science and Technology, 26*(3), 5159–5167.

The StaffofLab-Volt Ltd. (2006). RoboCIMSoftwareUser Guide37864-E0, LabVolt, Canada.

Vijayalaxmi, P. B., Rohan, P., Gayatri, S., & Punit, L. (2013). Object detection using image processing for an industrial robot. *International Journal of Advanced Computational Engineering and Networking, 1*(7).

Zahraa, D. H., Muhannad, Z. K., & Iman, S. K. (2015). Optimize path planning for medical robot in Iraqi hospitals. *Engineering Technology Journal, 33*(A), 5.

Chapter 5
Supportive Ecosystem for Leveraging Innovations and Promotion of Start-Ups and Entrepreneurship in India

Madhusudan Bandyopadhyay

Abstract The economic liberalization initiated by the Government of India in 1991 opened up tremendous opportunities for entrepreneurship and industrial development in the country. The government has carried out many policy reforms from time-to-time, depending upon the changing needs of society and the economy, and has started a number of schemes and programmes for the promotion of entrepreneurship and start-ups. Due to such progressive decisions, backed by a broad and diverse national innovation system, immense knowledge capital, and a favourable start-up ecosystem, the country is increasingly becoming a leading global innovation player, and has achieved an impressive growth in knowledge-intensive manufacturing and services. This paper describes various factors including the policies and programmes of the Government of India that have resulted in a supportive ecosystem for utilizing innovations, and the promotion of start-ups and new generation technopreneurs in the country. India's strengths in this regard, as well as its weaknesses in creating sustainable industries, have also been highlighted.

Keywords Ecosystem · Entrepreneurship · Innovation · Knowledge · Policy · Programmes · S&T · Schemes · Start-up · Support

Introduction

The term Innovation has a broad connotation, which refers to an idea perceived as new by the individuals. Thus, it relates to all spheres of life. In the context of industry, innovation can be regarded as the total process of the inception of an idea into the manufacture of a new or improved product or process, and finally of its commercial distribution in the market. Innovation is of importance *not only for increasing the*

M. Bandyopadhyay (✉)
Centre for Science and Technology of the Non-Aligned and Other Developing Countries (NAM S&T Centre), New Delhi, India
e-mail: namstcentre@gmail.com; bandodst@gmail.com

X. Poshiwa et al. (eds.), *Technological Innovations for Sustainable Industrialization*,
https://doi.org/10.1007/978-981-96-7995-9_5

wealth of nations, but also in the more fundamental sense of enabling men to do things which have never been done before (Freeman, 1974).

Technological innovation is a dynamic and complex process of converting a research result into a commercial enterprise leading to a positive change in the existing economy. Success of technological innovation always requires the existence of three factors, namely, Science and Technology (S&T) capability, market demand, and an agent, which transforms this capability into goods and services to satisfy the demand. An entrepreneur is that agent with his motivation and skills to carry out the process of innovation (Bandyopadhyay, 2018).

India has had a long and distinguished tradition in Science and Technology (S&T), which dates back to about 2500 BC. Application of technologies, innovation and entrepreneurship was an integral part of ancient Indian culture. According to archaeological evidence, people in India had skills in areas such as agriculture; astronomy and mathematics; medicine and surgery; metallurgy; town planning and architecture. About 200 sites of the Indus Valley Civilization are known today that exhibit the use of scientific knowledge and innovative practices.

India has the following strengths that have helped the country develop as a knowledge economy:

- Government policy support and funding;
- A broad and diversified National Innovation System with a vast S&T infrastructure, including institutions of excellence for higher education;
- A large high quality, English-speaking S&T manpower;
- Facilities for incubation;
- Global provider of IT and software services;
- A strong manufacturing base and dynamic private sector;
- Engagement of the diaspora.

Different ministries, departments, and agencies of the Government of India have taken many policy initiatives and implemented a number of schemes and programmes for the development and commercialization of technologies; providing training in entrepreneurship to S&T- qualified personnel; creating institution-industry linkages; and providing financial assistance and fiscal incentives to entrepreneurs. These initiatives have resulted in tremendous growth in start-ups and entrepreneurship in knowledge-based industries throughout the country. India is now regarded as a leading global innovation player.

This paper illustrates the success story of the promotion of innovation and entrepreneurship in India including the many schemes and programmes that have been launched by the Government for this purpose. The strengths and weaknesses in this regard, along with the barriers to sustainable industrialization in the country have also been discussed in the paper.

Major Factors for India's Success in Science, Technology, and Innovation

Policy Support for Research and Innovation

The Science & Technology (S&T) policy and strategy have evolved over the years since the Independence of India providing broad guidelines for scientific research and technological development. The first such policy was the Scientific Policy Resolution (SPR) (1958), then the Technology Policy Statement (TPS) (1983), the Science & Technology Policy (2003), and the Science, Technology and Innovation (STI) Policy (2013). Each time, the objectives and scope were redefined and expanded to address the country's changing needs over time, as well as economic liberalization and globalization in the 1990s. Currently, the process for formulating a new STI Policy, 2020 is going through several stages of consultations among various stakeholders (DST. https://dst.gov.in; Bandyopadhyay, 2023).

These policy documents emphasize, among others, the need to meet the development objectives of the country, including fostering innovation, industry and competitiveness. In its strategic documents, the government has recommended using international collaboration effectively to develop world class facilities in the country in selected areas of S&T and to enhance India's participation in major international S&T mega projects. It has proposed building strategic partnerships with other countries to facilitate a knowledge-based economy in selected industrial sectors. Most importantly, it has suggested that technology-led diplomacy should be used for enhancing India's image abroad, both from the geo-political perspective and for the development of S&T (DST, https://dst.gov.in).

The Science Advisory Council to the Prime Minister (SAC-PM), in its vision document titled "India as a Global Leader in Science: A Vision for India" stated: "*the potential for innovation is extremely high since diversity and adversity, both crucial to innovation, are characteristic features of the Indian scene......*" The report recommended various measures to create an ecosystem to translate India's potential into reality, including, venture funding and tax incentives, encouragement to scientists and technologists to turn their ideas into commercial ventures, extra-budgetary grants to government funded S&T institutions to cover risks associated with new ideas and innovation, encouraging tie-up among academia, R&D institutions and industry, strengthening IPR laws, and special package for promoting R&D intensive small firms (SAC-PM, 2010).

The Government declared 2010–2020 as the "Decade of Innovations". The main aim was to develop a favourable ecosystem in the country to stimulate innovations and utilize them for the societal good. In pursuance of these objectives, the Government of India initiated a number of programmes, which included financial support for technology development and innovation; promotion of start-ups and entrepreneurship; providing facilities for academia-industry collaboration; policy support to industry, and encouraging grassroots and inclusive innovations. Several institutional facilities were set up such as the National Innovation Council (NIC); the

National Science & Technology Entrepreneurship Development Board (NSTEDB), Science & Technology Entrepreneurs Parks (STEPs), Technology Business Incubators (TBIs), Atal Incubation Centres (AICs), the Biotechnology Industry Research Assistance Council (BIRAC), the National Innovation Foundation (NIF), and similar organizations. These initiatives resulted in an improvement in India's rank in the Global Innovation Index from 81st position in 2015 to an impressive 40th in 2023.

The Government has implemented appropriate changes from time to time in its approach and policies on S&T to encourage research and innovation that is relevant to the economy and society, promoting close and productive interaction between private and public S&T institutions, establishing an Intellectual Property Rights (IPR) regime, and creating a comprehensive national system of innovation covering science and technology, legal, financial and other related aspects.

In addition to the STI policies, separate policies and strategies have been adopted in various innovation-driven industrial sectors, such as information and communication, renewable energy, health, and food processing.

Robust National Innovation System

Since independence, a large infrastructure for research and innovation has been built in India with excellent facilities in various areas of science and technology. Autonomous research councils were set up to coordinate and fund research in different sectors. Concomitantly, an administrative support structure was created to plan, develop and administer S&T schemes and programmes in the country. Scientific ministries and departments, and S&T agencies in the public sector have established a large number of research and development (R&D) institutions. The higher education sector was strengthened by establishing more and more universities and institutions of national importance, and by introducing academic courses in modern disciplines of science, technology and engineering (Bandyopadhyay, 2023).

Over the years, a large number of R&D and higher educational institutions have been set up in the country. As of 2021, there were 7888 such institutions in India, with major contributors being: (i) the premier national R&D organizations, including Council of Scientific & Industrial Research (CSIR) (38), the Indian Council of Agricultural Research (ICAR) (103), the Indian Council of Medical Research (ICMR) (30), and the Defence Research & Development Organization (DRDO) (31); (ii) Ministries & departments—the Department of Science & Technology (DST) (21), the Department of Biotechnology (DBT) (16); and (iii) Higher Educational Institutions: Universities & Deemed Universities (993), Institutes of National Importance (127), among others (DST, 2021).

A recent important initiative of the Government of India was the establishment of the "Anusandhan National Research Foundation (ANRF)" in February 2024. The ANRF aims to provide high level strategic direction for research, innovation and entrepreneurship in the fields of natural sciences including mathematical sciences, engineering and technology, environmental and earth sciences, and health and agriculture. ANRF will seed, grow and promote R&D and foster a culture of research and innovation throughout the country's academic and R&D institutions by forging collaborations among the industry, academia, and government departments and research institutions. Established with a clear vision, the ANRF will play a crucial role in supporting and funding innovative projects across various disciplines (DST. https://dst.gov.in).

Strong and Vibrant S&T Manpower

India is well-placed with its excellent English-speaking S&T human resources and knowledge workers including the world's largest highly skilled IT workforce, as compared to other developing countries. According to an estimate for 2018, nearly 552,000 S&T personnel were employed in the country's R&D establishments in the country, and there were 255 researchers per million populations (DST, 2020). Further, there is a large and vibrant Indian S&T diaspora spread across the globe. The Government of India has initiated a few programmes in order to attract the diaspora to contribute to meet India's developmental challenges, and to promote India's international S&T outreach and cooperation in the countries where they live (DST, 2020).

Support for Start-Ups and Entrepreneurship

The Indian start-up ecosystem is the third-largest in the world currently with over 110 unicorns (Unicorn: a unit with valuation of more than US$ 1 billion) that have been setup in the last decade.

The expansion of start-up businesses and the development of an entrepreneurial culture gained momentum after the Government of India initiated economic liberalization in the 1990s, after which there was tremendous growth in the Information Technology (IT) and IT-enabled services industry. The Government has taken several policy initiatives that have given a major impetus to the start-up movement in the country. Start-ups in India have grown remarkably in the last few years and the number of recognized start-ups has increased to 117,254 as of 31st December 2023, as against 84,012 in 2022, and only 452 in 2016 (PIB https://pib.gov.in). These recognized start-ups are reported to have created over 1.242 million direct jobs creating significant economic impact. New generation entrepreneurs are at the forefront of introducing new technologies, products, and processes in order to scale up the growth of their businesses. These innovative and successful start-ups are driving value creation, employment generation and economic progress in the country.

In order to encourage the development of a favourable innovation and entrepreneurial ecosystem in the country, the Government of India, as well as various state governments, has implemented a number of policy measures and programmes. Such initiatives can be classified into the following groups based on their objectives:

(1) Creation of institution-industry linkages;
(2) Creation of institutional facilities—establishment of technology and business incubation facilities and accelerators;
(3) Providing support to innovators and entrepreneurs for the development and commercialization of technologies;
(4) Providing financial assistance on soft terms, and equity participation; fiscal incentives such as duty free imports, weighted tax deductions, tax holiday and income tax exemption and facilitating access to venture capital and angel investor funding;
(5) Skill upgradation and entrepreneurship development training, mentorship.

A number of institutions and funding mechanisms were set up to encourage innovation and start-ups. Under the Digital India Campaign, the country now has a digitally empowered society and has been transformed into a knowledge-based economy. The Government has launched several flagship programmes, such as the following, which have greatly advanced the promotion of innovation and industrialization, contributing to the overall economic development of the country:

1. **National Science and Technology Entrepreneurship Board (NSTEDB)**
   The National Science and Technology Entrepreneurship Board (NSTEDB) was set up in 1982 in DST for the promotion of technopreneurship and gainful self-employment among S&T personnel in the country. The Board has been supporting academic and research institutions to establish Science and Technology Entrepreneurs Parks (STEPs) and Technology Business Incubators (TBIs) in order to provide an institutional framework and extend support to technopreneurs. These programmes were perhaps attempted for the first time in the country primarily aimed at creating institution-industry linkages and providing entrepreneurship development training, technological services and business incubation facilities for the promotion of knowledge-driven and technology intensive enterprises (NSTEDB. https://www.nstedb.com).
2. **National Initiative for Developing and Harnessing Innovations (NIDHI)**
   Building on the experience gained from its activities, the NSTEDB started the National Initiative for Developing and Harnessing Innovations (NIDHI) in 2016, to nurture ideas and innovations (knowledge-based and technology-driven) into successful start-ups. The programme works in line with national priorities and goals, focussing on building an innovation-driven entrepreneurial ecosystem with the objective of socio-economic development through wealth and job creation. NIDHI aims to nurture start-ups through the scouting, supporting, and scaling up of innovations by providing a series of support components tailored to the critical initial phases of the start-up journey. Supports provided through various components of NIDHI are: for prototype development, to reduce risk for entrepreneurs, converting innovations to start-ups; for fast-tracking start-ups (accelerators), to provide seed support, and to help start-ups go global (NIDHI. https://nidhi.dst.gov.in).

3. **Development and Incubation of Technologies for Biotechnology Industry**

   The Biotechnology Industry Research Assistance Council (BIRAC) was set up by the Department of Biotechnology (DBT) as an interface agency to strengthen and empower emerging Biotech enterprises to undertake strategic research and innovation, addressing nationally relevant product development needs. Through its flagship programme, BioNEST, BIRAC has created 20 world-class bio-incubators with 200,000 Sq. ft. of incubation space which provide a multitude of support services, from office space to access to spend instrumentation and mentor networks. 65 companies have been established as start-ups through support from BIRAC (BIRAC. https://www.birac.nic.in).
4. **Promotion of Grassroots Innovations**

   India is a leader in grassroots innovations and there are a large number of community-based innovators with new ideas and possible solutions to local problems.

   The National Innovation Foundation (NIF) was set up as a national initiative for incubation and commercialization of technological ideas and innovations of grassroots innovators and students across the country. Its aim is to support grassroots innovators with institutional facilities and policies and transform the country into a creative and knowledge based economy. The NIF provides exposure and incubation facilities for the promotion of traditional knowledge, as well as indigenous and demand-driven innovations. Hundreds of grassroots innovations have been supported in a wide range of areas, such as technologies for the differently-abled, veterinary, plant varieties, agricultural-mechanization, textiles, food processing, and utility products (NIF. https://www.nif.org.in).

   NIF has set up the NIF Innovation and Entrepreneurship Council (NIF-ientreC) which is a Technology Business Incubator (TBI). It provides a comprehensive support system to provide timely help to innovators and entrepreneurs to develop and upgrade their ideas and innovations into commercially viable products and processes. It also offers them guidance and assistance for creating and operating a business venture and in disseminating innovations across India and abroad (NIF. https://www.nif.org.in).

   A Micro Venture Innovation Fund (MVIF) has also been created to extend financial support to grassroots innovators based on a single signature on a simple agreement of understanding, without any collateral or a guarantor. The key criterion for selecting a technology for MVIF support is the social return or value created for the benefit of society at large. It is a dedicated risk fund, established by NIF and supported by Small Industries Development Bank of India (SIDBI). The guiding principle of MVIF is timely and hassle-free financing of innovations (NIF. https://www.nif.org.in).
5. **Start-Up India**

   The Government launched the flagship initiative named, "Start-up India" in 2016 with the objective of building a strong ecosystem for nurturing innovation and start-ups in the country that would drive sustainable economic growth and generate large scale employment opportunities. The main components of the action plan of the Start-up India Programme are: (i) simplification and handholding, (ii) funding support and incentives, and (iii) industry-academia partnership and incubation. In order to provide funding support to start-ups, the Government had set up a fund with a total corpus of INRs 100 billion (~US$ 1.20 billion) over a period of four years (Start-up India. https://startupindia.gov.in).

Financial Support for Start-Ups

Under the Start-up India initiative, the Government has taken several steps to provide financial support for the development and growth of start-ups in the country. A few flagship schemes have been launched to provide support to start-ups at various stages of their business cycle to enable them to reach a stage where they can raise investments from angel investors or venture capitalists or seek loans from commercial banks or financial institutions.

The Funds for Start-ups (FFS) scheme was established in 2016 with a corpus of INR 10,000 crore (US $ 1200 mill), to provide much-needed boost to the Indian start-up ecosystem and enable access to domestic capital. The FFS, which is managed and operated by the Small Industries Development Bank of India (SIDBI), does not directly invest in start-ups; instead provides capital to SEBI-registered Alternative Investment Funds (AIFs), which in turn, invest money in growing Indian start-ups through equity and equity-linked instruments.

Another scheme, launched in 2021 is the Start-up India Seed Fund Scheme (SISFS), which provides financial assistance to start-ups for proof of concept, prototype development, product trials, market entry, and commercialization. The Seed Fund is disbursed to selected start-ups through eligible incubators across India. This support enables start-ups to reach a stage where they can raise investments from angel investors or venture capitalists.

In order to provide the much-needed collateral free debt funding to start-ups, and ensure a flow of financial assistance to innovative start-ups, the Credit Guarantee Scheme for Start-ups (CGSS) was launched. The scheme operates through a corpus provided by the government, which enables Non-Banking Financial Companies (NBFCs) to provide credit guarantees for loans extended to government- recognized start-ups by Scheduled Commercial Banks, NBFCs and Venture Debt Funds (VDFs) under SEBI registered AIFs. The scheme reinforces the Government's focus on promoting innovation and fostering entrepreneurship for making the Indian start-up ecosystem the best in the world and complements the existing schemes under the Start-up India initiative in mobilizing domestic capital for start-ups.

6. **Atal Innovation Mission (AIM)**

The Atal Innovation Mission (AIM) is a component of the Start-up India Programme, which was launched to promote a culture of innovation and entrepreneurship. It acts as a platform for promoting world-class innovation hubs, grand challenges, start-up businesses, and other self-employment activities, particularly in technology-driven areas.

AIM has two core functions: (i) Entrepreneurship promotion through Self-Employment and Talent Utilization (SETU), which is a Techno-Financial, Incubation and Facilitation Programme to support all aspects of start-up businesses, and other self-employment activities, particularly in technology-driven areas, and (ii) Innovation promotion: to support the establishment of new incubation centres called Atal Incubation Centres (AICs) that nurtures innovative start-up businesses in their pursuit to become scalable and sustainable enterprises.

AIM has successfully operationalized 72 AICs across India. These AICs enable start-ups by providing technical facilities, resource-based support, mentorship, funding support, partnerships and networking, co-working spaces and lab facilities, among other resources. Over 3500 start-ups have been incubated at these AICs and have created more than 32,000 jobs. AICs support start-ups from diverse areas such as Healthtech, Fintech, EdTech, Space and Drone Technology, AR/VR, Food Processing, Tourism, among others.

So far, under AIM 10,000 Atal Tinkering Labs have been established in schools across the country to foster curiosity and innovation in young minds, through advanced tools and technologies such as Internet of Things, 3D printing, rapid prototyping tools, robotics, miniaturized electronics, do-it-yourself kits and many more (AIM. https://aim.gov.in).

7. **Technology Innovation Hubs (TIH)**

Twenty-five (25) Technology Innovation Hubs (TIH) have been set up across the country through the National Mission on Interdisciplinary Cyber-Physical Systems (NM-ICPS). These hubs are boosting the development of new and emerging technologies, including Industrial Revolution 4.0 technologies, and are bringing out technological solutions for people-centric problems. The TIHs have also led to the establishment of 13 Technology Business Incubators (TBIs), 54 start-ups and spin-off companies, and the creation of around 928 jobs.

The NM-ICPS, which fosters technological solutions in health, education, energy, environment, agriculture, strategic security, and Industry 4.0, is being implemented through the TIHs established in top academic and national R&D institutes. The mission was launched with a total outlay of INRs 3660 crore (USD 440 million) in December 2018 (DST. https://www.dst.gov.in).

8. **SAMRIDH**

The Start-up Accelerators of MeitY for Product Innovation, Development and Growth (SAMRIDH) scheme was launched in 2021 to enhance the start-up accelerator ecosystem in India. It is designed to provide funding support to IT and software-based start-ups and help them integrate the necessary skill sets together to grow successfully (MeitY https: www.meity.gov.in).

9. **AGNIi**

Accelerating Growth of New India's Innovations (AGNIi) is a national initiative that aims to support the ongoing efforts to boost the innovation ecosystem in the country by connecting innovators in industry, the grassroots, and among individuals to the market and by helping commercialize innovative solutions (AGNIi https:www.agnii.gov.in).

10. **ASPIRE**

A Scheme for Promotion of Innovation, Rural Industries and Entrepreneurship (ASPIRE) was launched by the Ministry of Micro, Small and Medium Enterprises to establish a network of technology centres and incubation centres to accelerate entrepreneurship and to promote start-ups for innovation in agro-based industries.

11. **Start-Ups Intellectual Property Protection (SIPP) Scheme**

 The start-ups Intellectual Property Protection (SIPP) scheme was started in 2016 to encourage innovation and creativity in start-ups. Under this scheme, an 80% rebate on patent filing fees and a 50% rebate on trademark filing fees are provided to start-ups.

12. **Software Technology Parks of India (STPI)**

 Software Technology Parks of India (STPI) is a premier S&T organization under the Ministry of Electronics and Information Technology (MeitY) engaged in promoting IT/ITES industry, innovation, R&D, start-ups, and product/IP creation in the field of emerging technologies for various domains. So far, STPI has launched 24 Centres of Entrepreneurship (CoEs). Besides promoting R&D, innovation, and product and IPR creation, STPI is providing state-of-the-art infrastructure, skilling, mentoring, market connections, and other necessary support to start-ups throughout India. STPI has also embarked on launching a futuristic incubation scheme—the Next Generation Incubation Scheme (NGIS)—to offer comprehensive support and services and extend seed funding to start-ups from 12 STPI incubation facilities under a common umbrella. To further strengthen the start-up ecosystem, STPI has established advanced facilities and labs to enable start-ups to leverage these resources for developing innovative technology products and solutions (STPI. https://stpi.in).

13. **Make in India**

 The Programme aims to facilitate investment, foster innovation, enhance skill development, and build best-in-class manufacturing infrastructure, and encourage both domestic and foreign investments in the country.

14. **Stand-Up India**

 Launched in 2016, the Programme aims to promote entrepreneurship at the grassroots level for economic empowerment and job creation by facilitating access to Bank loans.

Contribution of States in India's Start-Up Ecosystem

Different states of India have contributed to the start-up movement in the country through dedicated policy initiatives. These initiatives promote start-ups and entrepreneurship by creating institutional facilities, such as incubators, innovation centres, and science parks; providing fiscal incentives and funding support; facilitating access to marketing, patent services, venture capital funding; and relaxing norms for government clearances. The initiatives of a few high performing states are discussed below:

Maharashtra

Maharashtra has an extensive start-up ecosystem and has the highest number of start-ups in the country, with over 25,500 entities. The state government has adopted the Maharashtra State Innovative Start-up Policy 2018, and initiated a number of schemes to encourage and support start-ups and entrepreneurs. These include self-certification for government clearances; funding and fiscal incentives, such as the Seed Money Scheme, District Industries Centre Loan Scheme, and Cloud space reimbursement; the creation of an investment fund for accelerators and incubators; preference for start-ups in state procurements; tax holiday, and assistance for quality testing and patent filing. The sectors in which start-ups have been established in Maharashtra include food and agriculture, education, skilling, healthcare, clean energy, environment, water and waste management, smart infrastructure, and mobility.

Karnataka

The Karnataka Start-up Policy 2015–2020 was launched with the aim of creating a thriving start-up ecosystem in the State and encouraging innovators and technology-based entrepreneurs. Programmes such as the ELEVATE initiative, the Grand Challenges Programme, and the New Age Innovation Network offer comprehensive support, including seed funding, venture capital, incubation and mentorship; networking opportunities, early stage funding, incentives such as tax exemptions, subsidies and reimbursements for patenting and marketing expenses, access to technology and R&D institutions, legal and financial assistance, training in entrepreneurship and business development; and self-certification for statutory approvals. With Bengaluru at its core, the state is a hotbed of innovation and entrepreneurial spirit, boasting over 15,000 start-ups across diverse sectors, such as information technology, biotechnology, and electronics system design and manufacturing (https://www.missionstartupkarnataka.org).

Tamil Nadu

Tamil Nadu is one of the high performing states with a favourable start-up ecosystem, boasting more than 120 incubators and 9000 start-ups. The state government has initiated a number of schemes and programmes to support start-ups, such as Start-up TN, TANSEED Start-up Capital Fund, and the SC/ST Industrial Fund. The State's skilled and talented population, coupled with an already industrialized economy and excellent academic institutions, contributes to its emergence as a start-up stronghold. The Tamil Nadu Start-up and Innovation Mission (TANSIM) supports start-ups in the state by promoting R&D investments; providing access to infrastructure facilities, knowledge creation, technology development and skill development.

To facilitate innovation through start-ups in the information technology and deep technology sectors, the state government has established the iTNT Hub (Tamil Nadu Technology Hub). Other sectors in which start-ups have been promoted in the state include agriculture, education, environment, ICT, healthcare, logistics, manufacturing, and urban development (Start-up TN. https://startuptn.in).

Sound Manufacturing Base in Innovation-Driven Industries

India has developed capabilities in a diverse range of industrial sectors. Due to its impressive R&D infrastructure, abundant and affordable skilled manpower, and a favourable climate for innovation, the country has become an attractive destination for investment from global multinational corporations and other large companies. These large industrial units provide immense scope for the growth of start-ups and small enterprises to serve them as vendors/suppliers of raw materials, components and spares; as ancillaries, business service providers, technology/software developers, consultants/advisors, distributors, or as downstream units.

The country has achieved impressive success in various innovation-driven industries. The salient features of some of these sectors are illustrated below.

Drugs and Pharmaceuticals Industry

India has become a global healthcare provider, and due to its huge manufacturing capabilities, with a large number of USFDA and WHO-GMP-compliant units, the country supplies affordable and low-cost generic drugs across the world. Due to its success, the Indian healthcare sector has earned excellent recognition from all over the world, which has led the country to be called "*The Pharmacy of the World*".

Currently, the Indian pharmaceutical industry is valued at INRs 4,20,000 crore (US$ 50 billion) (FY 2023–24) and is likely to expand to reach INRs 10,92,000 crore (US$ 130 billion) by 2030. India is the largest provider of generic drugs in the world. According to estimates, the industry supplies approximately 60% of the global demand for vaccines, 40% of generic demand in the US, and 25% of all prescription drugs in the UK, and over 80% of the anti-retroviral drugs used globally to combat AIDS (**Acquired Immuno-deficiency Syndrome**). The country played a commendable role in combating the COVID-19 pandemic by exporting life-saving medicines and medical equipment to over 150 countries. In 2021, Indian manufacturers exported more than 65 million doses of COVID vaccines to nearly 100 countries (Invest India. https://www.investindia.gov.in, IBEF https://www.ibef.org).

Information and Communication Technology (ICT) Industry

The Government of India has taken a series of policy initiatives to support the ICT industry, such as, the Digital India Programme (launched in 2015), the National Policy on Electronics (2019), and the National Broadband Mission (2019). It has created adequate digital infrastructure, provided a favourable legal and regulatory framework, and is enhancing rural connectivity. These steps have resulted in India emerging as a leading hub in the world for ICT-enabled and software services, accounting for about 55% share in the global outsourcing business. Contributing over 13% to India's GDP, the ICT sector is a major economic driver for the country; and with the expansion of the digital economy, the industry is expected to grow to about INRs 84 lakh crore ($1 trillion) by 2025. All subsectors of the ICT industry, including IT and business process management, IT-enabled services, hardware, software products, and e-commerce, have recorded double-digit growth.

In the recent past, India has developed significant capabilities in high-speed Internet connectivity, affordable smart devices and the development of software for digital applications and solutions. With its huge IT manpower, the country has become a global leader in new and emerging technologies associated with the Industrial Revolution 4.0, such as artificial intelligence (AI), big data, blockchain, genetic engineering, the Internet of things (IoT), nanotechnology, robotic process automation, machine learning, cloud computing, and cyber security (Invest India: https://www.investindia.gov.in, MeitY: https://www.meity.gov.in).

Renewable Energy Industry

India has made tremendous progress in the renewable energy (RE) sector, which is increasingly recognized internationally. The country ranks 4th globally in total RE installed capacity, with about 179 GW (as of July 2024). As part of its framework for transitioning towards clean energy to meet its commitments to achieve net zero, the country has significant potential for expansion in the RE sector, and thus offers a great opportunity for innovation-led investments. About INRs 17 lakh crore (US $ 200 billion) in investments have been made in new projects that are currently under implementation. The government is providing a number of policy supports and has set up RE/Solar parks to attract new entrepreneurs (Invest India: https://www.investindia.gov.in, MNRE: https://www.mnre.gov.in).

Biotechnology Industry

India's biotechnology sector is categorized into biopharmaceuticals, bio-agriculture, bio-IT, and bio-Services. The country is among the top 12 destinations in the world for bio-innovation and bio-manufacturing and is the third-largest in the Asia Pacific. Valued at INRs 11.5 lakh crore ($137 Bn) in 2023, the Indian bioeconomy Industry is targeted to reach INRs 12.6 lakh crore ($150 Bn) by 2025, and INRs 25.2 lakh crore ($300 Bn) by 2030. The Department of Biotechnology has funded 51 Biotech-KISAN (Biotech Krishi Innovation Science Application Network) hubs, connecting Indian farmers with the best scientists and institutions, to empower farmers, especially women farmers, with information on soil health, irrigation, and new agri-technologies. Forty-four hubs have been setup in 15 Agro-climatic zones of the country and are conducting activities in 169 districts (Invest India: https://www.investindia.gov.in, DBT: https://www.dbt.gov.in).

Discussion

India has achieved impressive success in the creation of S&T infrastructure and human resources, promotion of research and development, transfer, and adaptation of technologies for industrial applications. Indian capabilities in science and technology cover a diverse range of disciplines, such as advanced materials, agriculture, astronomy and astrophysics, biotechnology, chemicals and pharmaceuticals, information and communication technology, nuclear energy, oceanography, renewable energy and space technology, and applications. Modifications have been carried out in the S&T strategy, policy, and planning from time-to-time to meet the ever-changing developmental needs of the country.

The economic liberalization in India in the 1990s opened up huge opportunities for start-ups and entrepreneurship. Various policy measures have been implemented for the development of an innovation ecosystem and the promotion of entrepreneurship and knowledge-driven enterprises. The objectives of such programmes include the creation of institution-industry linkages, establishment of technology and business incubation facilities; providing support to innovators and S&T-based entrepreneurship development; development and commercialization of technologies; and providing financial assistance and fiscal incentives for the promotion of technology-based industries; establishing an Intellectual Property Rights (IPR) regime, and creating a comprehensive national system of innovation covering S&T, legal, financial and other related aspects. This multi-pronged approach has yielded favourable results with impressive success in different industrial, knowledge, and other skill oriented sectors, which have been driving the economic progress of the country.

The Government of India has provided an appropriate policy and regulatory framework and initiated a number of promotional measures and support services, leading to significant success in the advancement of industries in sectors like biotechnology, drugs and pharmaceuticals, ICT, renewable energy, etc.

India is increasingly becoming a leading global innovation hub. The country has achieved significant growth in recent years, even during the severe economic recession in other parts of the world. This growth is primarily due to progressive

Table 5.1 Global innovation index rankings of India

Year	GII rank
2016	66
2017	60
2018	57
2019	52
2020	48
2021	46
2022	40
2023	40

policies and innovative practices for developing capability in Science, Technology, and Innovation (STI) and knowledge-intensive manufacturing and services.

As shown in Table 5.1, India has made a consistent improvement in its innovation ranking in the Global Innovation Index (GII) published by the World Intellectual Property Organization (WIPO). This improvement is due to several policy initiatives, the country's immense knowledge capital, a supportive start-up ecosystem, and impressive progress made by public and private research organizations (World Intellectual Property Organization WIPO): www.wipo.int).

A few key takeaways of the 2023 Global Innovation Index report:

- India retained its 40th position for the second consecutive year.
- India ranks first in Central and South Asia.
- India leads the lower middle-income group of nations.
- India, along with China and Turkey, has consolidated its position as a global innovation powerhouse.
- India, along with Moldova and Vietnam, has outperformed expectations for the 13th year in a row.

According to the WIPO Report 2022, India is ranked 7th in the world for Resident Patent Filing activity. Further, India improved its ranking to 60th position (2023) in 2023 from 79th position in 2019 in the Network Readiness Index (NRI) 2023 report (The NRI is one of the leading global indices on the application and impact of ICT in 134 economies around the world).

Catalysing Sustainable Industrialization: Knowledge-intensive enterprises, particularly start-ups, play a crucial role in realizing the Sustainable Development Goals (SDGs). Start-ups have an inherent capacity to adopt new technologies and develop products and services that are sustainable in terms of environmental, social and economic aspects. Start-ups can play a major role in the implementation of SDG-9, which aims *to build resilient infrastructure, promote inclusive and sustainable industrialization, and foster innovation.* However, this potential of start-ups can be exploited only if the relevant policies and support mechanisms required for sustainable industries are available. For instance, performance on the two indicators of Target-9.5 [9.5.1: Research and development expenditure as a proportion of GDP, and 9.5.2: Researchers (in full-time equivalent) per million inhabitants] should be adequate to catalyse the creation of sustainable industries.

In this context, it should be mentioned that in spite of India's impressive success in innovation and entrepreneurship due to the proactive measures taken by the Government, the country still has a long way to go to create industries that are truly sustainable.

With regard to SDG-9.5.1, as described above, though India has made tremendous progress in STI with a strong S&T infrastructure and human resources, its Gross Expenditure on R&D (GERD) continues to be very low, estimated at only 0.64% of GDP (in 2020–21), which is significantly lower than that of industrially advanced countries, and even other BRICS countries. Moreover, India's GERD is driven mainly by public sector funding, and the contribution of the private sector was only 36.4% (in 2020–21), compared to that in developed countries (around 70%), and the global average (about 65%). Similarly, with regard to SDG-9.5.2, the number of researchers per million inhabitants in India was only 262 (in full-time equivalent) (in 2020) as compared to the world average of 1352 (2021). Thus, these two performance indicators for India are far from satisfactory in terms of the implementation of SDG-9.5.

Conclusion

The Government of India has created vast infrastructure for research and development (R&D) in various parts of the country which has resulted in impressive capabilities in different disciplines of science, technology and innovation. Many policy initiatives have been taken and a number of schemes and programmes implemented for the promotion and support for S&T based entrepreneurship development and commercialization of innovative technologies. These initiatives have resulted in tremendous growth in start-ups and knowledge-based industries throughout the country. Over the last decade, India has consistently improved its ranking in Global Innovation Index and is now regarded as a leading global innovation player.

However, in spite of such successes, the country still has to take corrective steps in order to create industries that can be regarded as really "*sustainable*". One of the major criteria for achieving the SDG-9 on innovation, entrepreneurship and industries, the country's investment on R&D is significantly lower than that of most other emerging economies. Moreover, the contribution of the private sector in the country's total R&D expenditure is very low indicating limited in-house innovation capabilities in the Indian industries. Further, another important target for SDG-9, the number of researchers per million populations in India is far from satisfactory. Therefore, in order to give further thrust to the start-up movement and improve sustainability of industries in the country, the Government must allocate increased funding for R&D, transfer of technologies and S&T manpower development, along with favorable policies, incentives, and subsidies to encourage private sector participation in R&D.

List of Abbreviations

AGNIi	Accelerating Growth of New India's Innovations
AI	Artificial Intelligence
AIC	Atal Innovation Centre
AIF	Alternative Investment Fund
AIM	Atal Innovation Mission
ANRF	Anusandhan National Research Foundation
ASPIRE	A Scheme for Promotion of Innovation, Rural Industry & Entrepreneurship
BIRAC	Biotechnology Industry Research Assistance Council
BRICS	Brazil, Russia, India, China, South Africa
CGSS	Credit Guarantee Scheme for Start-ups
CoE	Centre of Entrepreneurship
CSIR	Council of Scientific & Industrial Research
DBT	Department of Biotechnology
DRDO	Defence Research & Development Organization
DST	Department of Science & Technology
FFS	Fund of Funds for Start-ups
GDP	Gross Domestic Product
GERD	Gross Expenditure on R&D
GII	Global Innovation Index
GMP	Good Manufacturing Practices
ICAR	Indian Council of Agricultural Research
ICMR	Indian Council of Medical Research
ICT	Information & Communication Technology
IoT	Internet of Things
IPR	Intellectual Property Rights
IT	Information Technology
MVIF	Micro Venture Innovation Fund
NBFC	Non-Banking Financial Company
NGIS	Next Generation Incubation Scheme
NIC	National Innovation Centre
NIDHI	National Initiative for Developing and Harnessing Innovations
NIF	National Innovation Foundation
NM-ICPS	National Mission on Interdisciplinary Cyber-Physical Systems
NRI	Network Readiness Index
NSTEDB	National Science and Technology Entrepreneurship Development Board
R&D	Research and Development
RE	Renewable Energy
S&T	Science and Technology
SAC-PM	Science Advisory Council to the Prime Minister

SAMRIDH	Start-up Accelerators of MeitY for Product Innovation, Development and Growth
SC/ST	Scheduled Castes/Scheduled Tribes
SDG	Sustainable Development Goal
SETU	Self Employment & Talent Utilization
SIDBI	Small Industries Development Bank of India
SIPP	Start-ups Intellectual Property Protection
SISFS	Start-up India Seed Fund Scheme
SPR	Scientific Policy Resolution
STEP	Science & Technology Entrepreneurs Parks
STI	Science, Technology and Innovation
STPI	Software Technology Parks of India
TANSIM	Tamil Nadu Start-up and Innovation Mission
TBI	Technology Business Incubators
TIH	Technology Innovation Hub
TN	Tamil Nadu
TPS	Technology Policy Statement
USFDA	United States Food & Drugs Administration
VDF	Venture Debt Fund
WHO	World Health Organization
WIPO	World Intellectual Property Organization

References

AGNIi. https:www.agnii.gov.in. Accelerating Growth of New India's Innovations (AGNIi). Government of India.

AIM. https://aim.gov.in. Atal Innovation Mission, Government of India.

Bandyopadhyay, M. (2023). Development paradigm of India as an enabler for practicing science, technology and innovation diplomacy. In Ittekkot, V., & Bawejs, J. K. (Eds.), *Science, Technology and Innovation Diplomacy: Perceptions and Practice*. Research for Development. Springer. https://doi.org/10.1007/978-981-19.6802-08.

Bandyopadhyay, M. (2018). Mechanisms for commercialisation of technology: The role of technology business incubators in the promotion of knowledge-based industries. In Al Akhzami, S. (Ed.), *Technology transfer and commercialisation*. Astral Publications, New Delhi.

BIRAC. https://www.birac.nic.in. Biotechnology Industry Research Assistance Council, Department of Biotechnology, Government of India.

DST. (2020). Research & Development Statistics 2019–20. Department of Science & Technology, Government of India, December 2020. https://www.nstmis-dst.org.

DST. (2021). Directory of R&D Institutions. (2021). Department of Science & Technology, Government of India. https://www.dst.gov.in

DST. https://dst.gov.in. Department of Science and Technology, Government of India.

Freeman, C. (1974). *The economics of industrial innovation* (2nd ed.). Frances Printers.

Mission Start-up Karnataka. https://www.missionstartupkarnataka.org

IBEF. https://www.ibef.org. India Brand Equity Foundation.

Invest India. https://www.investindia.gov.in.

MeitY. https:www.meity.gov.in. Ministry of Electronics and Information Technology, Government of India.

NIDHI. https://nidhi.dst.gov.in. National Initiative for Developing and Harnessing Innovations, NSTEDB, Department of Science and Technology, Government of India.

NIF. https://www.nif.org.in. National Innovation Foundation.

NSTEDB, https://www.nstedb.com, National S&T Entrepreneurship Development Board (NSTEDB), Department of Science and Technology, Government of India.

SAC-PM. (2010). India as a Global Leader in Science: A Vision for India. Working Papers id:2940, Science Advisory Council to Prime Minister.

Start-up India. https://startupindia.gov.in. Start-up India, Government of India.

Start-up TN. https://startuptn.in.

STPI. https://stpi.in. Software Technology Parks of India, Ministry of Electronics and Information Technology, Government of India.

World Intellectual Property Organization (WIPO). (2023). *Global Innovation Index 2023: Innovation in the face of uncertainty*. WIPO. https://doi.org/10.34667/tind.48220

Chapter 6
Articulating the Relationship Between Research, Innovation and Industrialisation in the Context of Education 5.0-A Perspective

Isaac Nyambiya

Abstract The principle and practice of innovation underpins today's global economies which are challenged by the dichotomy of unbridled consumerism and wastage in the Global North and chronic malnutrition in the south. The global nature of the COVID-19 pandemic, however, has shown that this divide is unsustainable and that an effective global response requires research and innovation by the citizenry. Western countries seem to have mastered the pillars of production through the generation of intellectual property monetised into processes and products that benefit society. State institutions of higher learning (IHL) and research centres in Africa largely depend on public funding and are struggling to fulfil the mandate of generating and maintaining knowledge whilst bringing forth innovations which make them self-sustaining. As such, IHL are limited in their significant contribution to national development apart from churning out students for the workplace who do not become successful innovators and entrepreneurs. Part of the problem lies in the inability of faculty members to conceptualise and articulate the concepts of research and innovation and their relation to the IHLs themselves and the wider economy. This affects execution. This paper seeks to explore the connection between research and innovation which results in enterprise in the context of Education 5.0. The aim is to help bridge the interface between higher education and industry, which has always been difficult to navigate, insofar as industry may only partner with tertiary institutions in IP generation and partnerships, if and only if, they see a material benefit which enhances their own interests. Universities in Africa can move away from dependence on government funding through actively cultivating symbiotic partnerships with industry, development agencies, and civic organisations on cross-platform initiatives which are mutually beneficial.

I. Nyambiya (✉)
Department of Physics, Geography and Environmental Science, Faculty of Science, Great Zimbabwe University, Masvingo, Zimbabwe
e-mail: inyambiya@gzu.ac.zw

X. Poshiwa et al. (eds.), *Technological Innovations for Sustainable Industrialization*,
https://doi.org/10.1007/978-981-96-7995-9_6

Keywords Innovation · Education 5.0 · Technology · Intellectual property · Entrepreneurship · COVID-19

Introduction

Questions have been asked as to whether the traditional roles of the university are sustainable in the future in an era of greater scrutiny, high competitive resource allocation, and pressure for creativity to attract new income streams in the information age (Scott et al., 2008). This became more apparent during the COVID-19 pandemic when universities were called upon to offer solutions to mitigate and end the crisis (Aristovnik et al., 2020; Daniels, 2020). Over the past 30 years or so, the functions of universities have thus been evolving to inculcate the 'entrepreneurial university' as a change-agent and solutions agenda (Bhayani, 2015).

To this end, Zimbabwe has embarked on a path which deliberately incorporates innovation and industrialisation (called Education 5.0) as key deliverables in institutions of higher learning (IHL) moving away from the self-limiting traditional university (Education 3.0). However, in order for these pillars to be embraced by faculty it is imperative that there is an unambiguous understanding of the parameters involved in embedding and engendering sustainable innovation which benefits society. The word innovation has become ubiquitous in modern language to characterise the digital economy underpinned by the Fourth Industrial Revolution. Predictions have been made that by 2020, between 26 and 200 billion people and things would be connected to networks (International Telecommunication Union, 2015). However, innovation goes beyond just connectivity as understood in the modern sense (Pticek et al., 2016) even though connectedness today has become one of the biggest drivers of national development (Mačiulis et al., 2009).

Governments in Africa are aware of the need for IHLs to go beyond just generating and maintaining knowledge, but to also be innovative in order to sustain themselves and significantly contribute to the fiscus and national development (Ministry of Higher and Tertiary Education Innovation Science and Technology Developments, 2018). The introduction of Education 5.0 in Zimbabwe has added innovation and industrialisation as additional pillars to the deliverables of teaching, research and community service (Muzira & Bondai, 2020).

Researchers, however, observe the dearth of a clear perspective in understanding the role of higher education in the innovative process (Uyarra, 2008) because of the lack of universal best practices for University-Industry-Collaborations (UIC) (Awasthy et al., 2020). Articulating the link between research and innovation is therefore important to establish its connection to the various pillars of development in order to guide execution within institutions.

Literature Review

The connection among research, innovation and industrialisation is demonstrated by the body of literature dealing with this subject (Barasa et al., 2017; Dessler, 1969; Martinuzzi et al., 2018; Miguel Benavente, 2006; Nicolaides, 2014; Thapa et al., 2019). The link appears simple: scientific research leads to innovation (Celeste et al., 2014), which ultimately leads to industrialisation and national development. Smaller countries such as Singapore which have very few natural resources to their credit have exemplified how this nexus can be executed by increasing their per capita income from US$516 in 1965 to US$52,888 in 2015 (Poh, 2016). Natural resource-deficient countries such as Germany have survived as technology-driven nations because of the priority placed on innovations (Kinkel et al., 2005).

Using a systems approach, Godin (2010) describes the interaction between universities, governments, and industry and not-for-profit organisations in an ecosystem that creates innovations. Practically navigating the bridge between research and innovation, and thus providing a vehicle for national development, is what has been problematic in African IHL (Diop, 2017; Murgor, 2015; Sawyerr, 2017). Using one's intellectual prowess to implement the imperatives that transform tangible and intangible resources, even ones outside a country's jurisdiction, is what defines innovation. However, researchers have argued that innovation is more than research and development (Kinkel et al., 2005), and to a large extent, Africa faces the challenge of a lack of a well-articulated and supported National System of Innovation (Bartels et al., 2016).

At the very basic level, the classical trajectory of innovation is encapsulated in the movement from *imitation, invention* and finally *innovation* (Godin, 2008). Imitation conjures apprehension when mentioned in academia, as this implies copying or plagiarism. However, copying can also come in the form of imitating and simulating nature through creativity, resulting in inventions that transform into innovations.

Defining Research and Innovation

Naidoo gives an all-encompassing definition of research, describing it as "the diligent systematic enquiry into nature and society to validate and refine existing knowledge and to generate new knowledge" (Naidoo, 2011). Meanwhile, the term innovation has evolved from its genesis in the arts through a disdain in religious systems, engineering of social order, and the introduction of technologies in the eighteenth century (Godin, 2008). Today, innovation retains a multitude of meanings, as demonstrated in the philosophy which see this process as encompassing all the scientific, technological, organisational, financial and commercial activities necessary to create, implement, and market new or improved products or processes (Léger & Swaminathan, 2007).

However, the simplest definition of innovation remains "the provision of a solution to a problem" (Taylor, 2017). J.A. Schumpeter is credited with articulating the concepts of the five types of innovation, namely the *introduction of a new good, the introduction of a new method or process of production, the opening of a new market, the conquest of a new source of raw materials,* and *the implementation of a new form of organisation* (Godin, 2008). To this extent, an invention is not an innovation until it has an *utilitarian value* beyond the inventor (Stock, 2014).

Human Resource

Resources are recognised as the heritage through which innovation is achieved. Central to the concept of innovation is the existence of a key human resource endowed with creative thinking to activate solutions at any level of society. Development practitioners note that the most critical driver for innovation is the quality of human resource within and without the organisation (Banerjee, 2014). Innovation is directly linked to the extent, quality and variety of education and skills of the citizenry (Scott & Vincent-Lancrin, 2014). As such, scholars have associated the trajectory of innovation with immigration and emigration, the former enriching innovation (Maré et al., 2011), whilst the latter depletes the critical human capital important for advancing the sciences, literature, the arts, industry, law and religion, as was the case in France for a period of about 300 years during the Dark Ages (White, 1888).

Centrality of Natural Resources

Also critical to the discourse on self-sufficiency of institutions and nations is the legitimate ownership of resources, including intellectual capital. Other natural endowments, in the form of abundant biomass (Bar-On et al., 2018), water (Goonetilleke & Vithanage, 2017), minerals (Dubiński, 2013), the sun and wind (Sivaram et al., 2018; Vidican, 2015), have been recognised as the basis upon which research is practised, applied, and transformed into innovations (Manley, 2002). In her book The Ministry of Healing, Ellen G White reveals part of the reason why the innovations which the world has garnered over centuries, do not seem to provide the transformative disruption intended to eliminate hunger and poverty, writing:

In God's plan for Israel, every family had a home on the land, with sufficient ground for tilling. Thus, both the means and the incentive were provided for a useful, industrious, and self-supporting life. And no human devising has ever improved upon that plan. To the world's departure from it is owing, to a large degree, the poverty and wretchedness that exist today (White, 1905).

The above statement posits the philosophy that to reduce hunger and poverty, equitable distribution of land is paramount for development, without which a skewed form of industrialisation follows, enriching one group of people at the expense of another. Other authors corroborate this view and note that no nation has achieved sustained growth without addressing issues of food security, putting land and agriculture at the centre of innovation and economic development (Juma, 2011). At one point in its development, the USA was an agrarian economy (Allison et al., 1976; Schmidt, 2021) before transitioning to a service-based economy (Ginzberg & Vojta, 1981). The rapid development of China is underpinned by agriculture, where about 200 million small-scale farmers, each possessing about 0.6 hectares of land, support a population of 1.3 billion people (Juma, 2011).

Models of Innovation

Models of innovation describe the innovative process from both historical and exploratory approaches, presenting a mental picture of how the development of innovation and its functions have evolved over time.

The linear model of innovation is understood as the first representation articulating the connection between science, technology and the economy. The model evolved over the years, incorporating input from the natural scientists, industrialists, business people and ultimately economists who appeared much later after the science (Godin, 2006). Also called the ladder model, its pre-eminence derives from the "success" of the atomic bomb and other technologies that demonstrated how the dominance of science led to technological prowess (Mahdjoubi, 1997). This model has shaped scientific thinking for over 200 years (Godin, 2006).

The nonlinear models are alternative representations that critique the weaknesses of the ladder model, such as describing the process of development as being framed in discrete components operating indepen dentlyof each other (Mahdjoubi, 1997). The nonlinear models add feedback loops to the innovative process, which is treated as a whole (Léger & Swaminathan, 2007).

The circle model developed by IBM's Ralph Gomory is viewed as representing a second perspective on the relationship between science and technology. Its proponents argue that the circle model is not science-based and recognises that innovation can originate from anywhere (Mahdjoubi, 1997). Key to the innovative process is evolution, characterised by iteration, resulting in various generations of the same products. The chain-linked model envisions innovation as an intricate whole with feedback loops that inform the entire process. Visually, it presents research as an integration bar linking knowledge with practice/enterprise (Léger & Swaminathan, 2007).

Theories of Innovation

The economic theories of innovation include institutional economics, industrial economics and evolutionary economics of international trade. Central to these theories are issues of intellectual property rights and the endogenous factors that affect the process of innovation, whilst international trade involves the external circumstances contributing to firm level innovation (Léger & Swaminathan, 2007).

National System of Innovation

In order to stimulate the process of innovation, many countries around the world, including those in Africa, have instituted National Systems of Innovation (NSI) (Bartels et al., 2016). NSIs are policy blueprints meant to harness and guide intellectual prowess for national development. These NSIs are well-discussed, encompassing Agricultural Innovation System (AIS) (Agwu et al., 2008; World Bank, 2012), Education Innovation Systems (EIS) (Brennan et al. 2014; Cai et al., 2020), Mining Innovation Systems (MIS) (Bas & Kunc, 2009), Manufacturing Innovation Systems (MIS) (Bas & Kunc, 2009), Health Innovation System (HIS) (Berger et al., 2009; Consoli & Mina, 2009). IHLs have participated in formulating these policies, including their implementation (Gunasekara, 2006). Basically, the framework for NSI is one that describes research as a tool to achieve innovation in a system where the university uses knowledge as it interacts with governance, political systems and the economy to generate wealth (Godin, 2007). This is done in a collaborative effort that acknowledges the whole and not the discrete parts of the ecosystem. The aim of this paper is to articulate the stages in the innovation process which leads to industrialisation using a case-study approach and purposive document analysis.

Methodology

A three-tier mixed methods approach was used to investigate the understanding of innovation in Zimbabwe and to describe and explain the concepts of research, innovation and industrialisation through the lens of Education 5.0 among IHL faculty members in Zimbabwe.

General Google Search

To critically analyse the topic that connects research to innovation, industrialisation and enterprise, a general Google search was performed to randomly select articles that fit the historical development. Initially, a historical and narrative review was conducted to gain an overall understanding of the connections between the various topics, with a view to revealing the trajectory of how developed nations have reached their current state. This information is used throughout the paper to articulate and discuss the relationship between research, innovation and industrialisation.

Google Scholar on Education 5.0

A Google Scholar literature search of peer-reviewed publications on education 5.0 in Zimbabwe that were published between 1980 and 2022 was performed with a view to precisely including articles that specifically studied the direction of education in Zimbabwe. During this 42-year period, the Zimbabwean education system has transitioned from a racially segregated paradigm into an inclusive system of social transformation (Mungazi, 1985). The Google Scholar search engine was used to perform a keyword search using the following phrase: 'Education 5.0 in Zimbabwe'. The search yielded 113 papers, which were carefully screened to only include those articles considered relevant for this topic. Forty-eight papers were found to include the keywords 'Education 5.0' and 'Zimbabwe'.

Case-Study Literature

Furthermore, a case study of a university and its incubation holding company in the Global North was chosen and analysed as a success story of an institution that has successfully fostered an entrepreneurial university. Through this review, the author aims to encourage more rigorous debate in the Global South on articulating and implementing the connection between research and innovation whilst creating a platform for achieving industrialisation.

Results and Discussion

Perceptions of Education 5.0

Through its National Development Strategy 1 (Republic of Zimbabwe, 2021), the Government of Zimbabwe introduced the policy 'Doctrine for Modernisation and Industrialisation of Zimbabwe through Education, Science and Technology Development to achieve Vision 2030' (Ministry of Higher and Tertiary Education Innovation Science and Technology Development, 2019). This vision articulated by the government is intended to transform the country into a 'prosperous and empowered upper middle income society by 2030'. The vehicle envisaged for achieving this is the education system, particularly IHLs, which have been tasked as agents that drive innovation and industrialisation beyond the traditional roles of teaching, research and community service. This philosophical shift has been imposed particularly on lecturers actively involved in the pursuit and delivery of knowledge to transform society. At the highest level, educators recognise the increasing redundancy of IHLs in Zimbabwe and the need to reform in an environment where resources from central government are dwindling (Zvobgo, 2019). University educators and administrators have recognised the stagnation in resource generation that has affected IHLs for over 30 years (Chinyoka & Mutambara, 2020).

A study of the literature published on Education 5.0 in Zimbabwe emphasises the need for research that leads to economic development. There is a tacit realisation to change the status quo of the country through a shift that entrenches the production of goods and services to benefit society through a knowledge-based economy. In 2017, the Research Council of Zimbabwe held its 11th Zimbabwe International Research Symposium with the theme 'Research for Unlocking Entrepreneurship: Bridging the Gap' (Research Council of Zimbabwe, 2017). This seems to show the collective struggle within the IHLs to break with the past and begin to impact the economy and society in tangible ways that eclipse the ivory tower status that the IHLs have maintained for a very long time.

The Education Blueprint which highlighted the teaching of STEM subjects at primary and secondary levels was introduced in 2016 (Chitate, 2018). The enthusiasm around the policy seems to have disappeared because of a change in government that has introduced a similar policy of Education 5.0 aimed at institutions of higher learning even though some in the IHL argue that this approach is not really not new (Keche, 2021). However, the fact that educators appear to be worried about how the introduction and implementation of Education 5.0 is being undertaken (Muzira & Muzira, 2020) seems to suggest that there has not been a clear articulation of the government's new thrust in spite of its good intentions.

Recent studies on the topic have shown that educators at IHL in Zimbabwe are not averse to the implementation of the philosophy of Education 5.0. They urge the government and the universities to improve infrastructure and increase funding to help in meeting the additional pillars of innovation and industrialisation (Muzira & Bondai, 2020). There is also a realisation among faculty that there is a strong link between the knowledge hub (the university) and a product incubator as a way of creating tangible value out of products and processes from this ecosystem (Murairwa, 2021). Most state universities in Zimbabwe, by government policy, have established innovation hubs that are meant to commercialise and monetise the intellectual property generated from their research. Additionally, universities in Zimbabwe have also been teaching courses on innovation and entrepreneurship as disciplines (Musindo & Mudavanhu, 2006). Globally, research has shown a definitive improvement in the economy as a function of the teaching of entrepreneurship. Zimbabwe's case seems to trend in the opposite direction, pointing to wider and deeper systemic issues within the education and economic ecosystem in the country (Jengeta, 2020).

Researchers in Zimbabwe have also recommended the entrenchment of intellectual property within the philosophy of Education 5.0 as part of the policy framework that guides and incentivises the genuine adoption of innovative practice among faculty (Bigirimana, 2021). However, records show that the highest number of patents to date, numbering over 300, were filed in 1980. Since then, there has been a decline to fewer than 20 per year. One study also noted that from 1996 to 2016, although the number of published articles at one Zimbabwean engineering school increased, the number of patents, which can be used as a proxy for the capacity to industrialise, dropped by about 6.1-fold during the same period (Mamina & Maganga, 2019).

Some researchers have also pointed to the need to infuse creativity and critical thinking as part of the skills taught to engender innovation, encouraging the inclusion of arts education, which is presumably known to foster these skills (Dziwa & Postma, 2020). Others have noted the need to capacitate faculty leaders and improve leadership styles while implementing appropriate staff-student ratios that do not overburden the faculty (Guvhu & Museva, 2020). A case for the use of indigenous languages in IHL has been made, arguing that students will have a better capacity for innovation if their learning is conducted in local languages (Muyambo & Ndamba, 2021).

Whilst the lecturers and educators at the ground level are not averse to the introduction and implementation of Education 5.0, there is a strong sentiment that such drastic changes should happen after thorough consultation with the faculty concerned to guarantee sufficient buy-in (Kaviya et al., 2022).

Case Study

Some universities in the Global North have made significant progress in entrenching innovation and industrialisation that is worth emulating. The following is an analysis of university A in the Global North, which has fostered research, innovation and industrialisation among its faculty.

The university boasts more than a century of existence since its founding as college in 1845 and achieving university status in 1908, with more than 20 schools and 11 specialised institutes (Universities UK, 2015). The university has had a considerable history of academic entrepreneurship since the formation of a holding company in 1984 as a commercial and incubation arm of the university involved in technology transfer through linking academic and industry entrepreneurs (QUBIS, 2015a). With the slogan 'from research to wealth', the incubation company has helped to 'spin out' more than 90 companies (QUBIS, 2015b). The spin-off companies formed through the company were, as of 2016, contributing as much as GBP120 million per year to the Northern Irish economy (QUBIS Ltd Belfast, 2016).

Figure 6.1 below represents the trajectory of the research output at the university, as far as documented records can show. Over a 50-year period, the university has accumulated a repository of at least 96,457 publications. Figure 6.2 shows the extent of the types of publications. The records show more than 46 types of publications ranging from 'articles' and 'conference contributions', to 'abstracts', 'films', 'protocols', 'entries for encyclopedias/dictionaries', 'patents', 'software', etc. Slightly over 50 years ago (1973), the university was producing as few as seven publications per year. By the time the incubation company was formed in 1984, the research output had increased more than 14-fold to 114 publications, which represented a significant percentage improvement. Research output has been increasing almost exponentially to a peak of 5268 publications in 2016. However, there has been a decline over the last five years, resulting in a difference of about 1000 published articles between 2016 and 2020 (QUB, 2022).

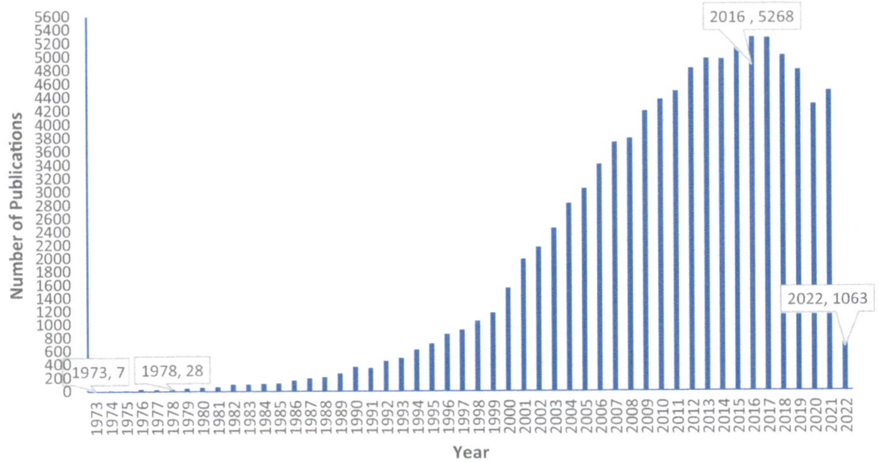

Fig. 6.1 Number of publications at university A since 1973 until 2022

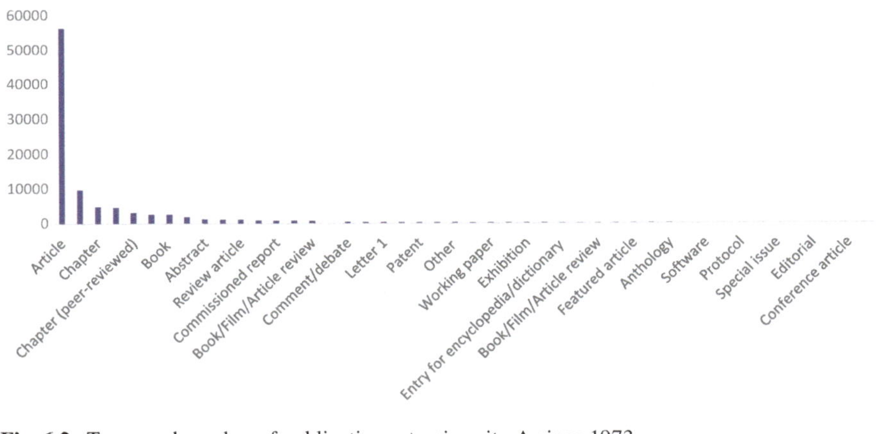

Fig. 6.2 Types and number of publications at university A since 1973

By far, the largest number of publications has been the combined output of 'articles' at 54.8%. Very curiously though, out of the many publications produced, only 241 were patents, were produced over the same 50-yeartime period, yielding a patent publication rate of 0.25%. Patents are often cited as an index for a nation's industrial development (Griliches, 1990). However, through the incubation company incorporated to commercialise the university's research output, stemming from the number and quality of patentable research, the university is considered one of the top institutions in Europe in this regard.

Predominantly, English is used as the language of publication, with 94,679 publications (98.2%) output in English. However, at least 14 other languages have been used to publish work from the university. Collaboration with other authors has occurred with at least 50 universities and institutions, mainly in the UK but extending globally.

Research in most African universities has been described as a 'lost mission' for over four decades because of a myriad of circumstances, such as restrictive political environments (Andoh, 2017). However, a number of points for discussion can be raised to foster better implementation.

As of 2021, Zimbabwe ranked 113th out of 132 countries on the Global Innovation Index, 24th out of 34 middle income countries and 13th out of 27 Sub-Saharan countries in the region. The data shows that Zimbabwe's weakest performance is in its institutions, ranking 129th out of 132. What seems to stifle creative outcomes is the political environment (government effectiveness, regulatory quality, and the rule of law), which ranked at 131st out of 132 for Zimbabwe (WIPO, 2021). Because of this, Zimbabwe has suffered a crippling brain drain, with high-calibre researchers exiting the country over more than 40 years since the country gained independence nation without war (Chetsanga, 2003). For instance, skills audits have revealed a skills gap of 95% in the medical sciences, 93.57% for the engineering and technology sector and 96.91% for the natural and applied science cluster and 88% in the agricultural sector (Robinson, 2020). As such, the incapacitation of the local education system, which has happened progressively and decisively over the last 20 years, has contributed significantly to the inability of institutions to generate new knowledge.

Recommendation

University-Industry-Collaboration

Several models are recommended here for the African university to become fully invested in creating value for national development. The University-Industry-Collaboration (UIC) model has been practiced globally, with a rather late realisation of its importance in Africa (Ankrah & Al-Tabbaa, 2018). Albulescu and Albulescu state that a university with a strong entrepreneurial spirit maintains a strong relationship with the community from which it derives its existence (Albulescu & Albulescu, 2014). Researchers acknowledge that it is still difficult for organisations and universities to find universal best practices to follow in order to establish functional relationships (Awasthy et al., 2020), probably due to a lack of dedicated faculty with enough drive and mettle to formulate, maintain and sustain these relationships based on performance and perhaps charisma.

The recent phenomenon of the Innovation Hub in Zimbabwean universities, instituted by an Act of Parliament, is well-suited to play an interfacing role between faculty and industry, ensuring that a research and innovation-centric approach is cultivated on both sides for mutual benefit. Academic leadership in research output will remain paramount to attract well-resourced industry partners and retain their interest in funding mutual projects. Successful collaborations have been established on the continent, for example, in Nigeria through the Nestle and University of Agriculture, Abeokuta partnerships (Juma, 2011).

Open Innovation

One approach that can be employed is the principle of open innovation. Open innovation is the principle that accepts that the best ideas can come from anywhere, not necessarily from within a single organisation (Xiangjiang et al., 2013). Open innovation also accepts that breakthroughs in innovation achieved by one company can be shared with both competitors and customers, usually under license, to accelerate the innovation process for all parties (Xiangjiang et al., 2013).

Intellectual Property

The generation of original intellectual property remains the cornerstone of ownership of the products and processes of innovation. This ownership is allocated on the basis of a product or process meeting the conditions of patentability, which are utility, novelty, non-obviousness (inventive step), and being patentable subject matter within the jurisdiction of a nation or globally (WIPO, 2014). In the university setting, it is the practice of research that generates intellectual property upon which innovations are based. Universities can leverage their research capacity to research to collaborate with industry which can capacitate the university in its efforts to produce intellectual property and innovations (Abramo & Costa, 2011).

A Self-Sustaining Entrepreneurial University

State universities particularly in Zimbabwe face a plethora of problems, including inflationary pressures and reduction of government support, among other impediments (Chinyoka & Mutambara, 2020). Robust arguments have been proffered for (Harthy, 2014) and against the entrepreneurial university (Armbruster, 2008) as a panacea for weaning the university offits limiting dependence on diminishing government funding and support. The self-sustaining university is a principle that most state-owned IHL can adopt to achieve the requirements of Education 5.0 in the African context. The entrepreneurial university is a new conception of this self-sustaining institution (Samadi et al., 2018), capable of commodifying and commercialising its knowledge, even though this should not supersede the original mandate of the university as a creator of knowledge for its own sake. The problem of finding appropriate leadership to make this transition is one that has been identified globally (Gibb et al., 2009). Faculty has itself identified key tenets through which the university must operate to realise this model of development. These include, but are not limited to research and development while integrating teaching, establishment of growth hubs, industrial placements, entrepreneurship gap years, guest lectures and campus outreach programmes on entrepreneurship (Pugh et al., 2018).

Dangers of Innovation

There are existential problems associated with the blanket use of innovation, especially when innovation is treated in silos without considering the whole system. A few examples will make this point clear: Weapons of mass destruction have been developed for a long time period (Molella & Kargon, 2004). The Manhattan Project, which created the first atomic bomb, was a deliberate effort to use knowledge accumulated over centuries to inflict maximum damage, showing how science-based innovation as a competitive pursuit for world domination can be abused (Fehner & Gosling, 2012).

A reductionist approach, especially in the area of agriculture and nutrition, has brought more harm than good, even though society has often treated this merely as an innovation. For instance, the refining of sugar cane into a crystalline white solid may be considered an innovation because of the ease with which populations now enjoy sweetened products. However, this has come at the expense of health, as obesity, linked to free sugars, was declared a pandemic by the WHO at its inception in 1948 (James, 2008). Obesity is a risk factor for many of the cardiovascular diseases, which are among the largest causes of death globally, and this can be traced back to unbridled 'innovations'. While innovation is regarded as progress in making food readily, it has also increased the global burden of disease (Afshin et al., 2019).

The thalidomide tragedy, in which children were born without limbs after mothers took the drug to treat morning sickness, is another clear example of dangers associated with innovation (Kim & Scialli, 2011). As such, research should be accompanied by sufficient risk analysis before deployment. Whilst faculty are involved in the trajectory of research, innovation and industrialisation, it is equally important for them to flag potential deleterious issues associated with the research they conduct. As such, ethical and governance considerations are equally important to address during the transition in which Zimbabwe's IHLs are involved.

Conclusion

Whilst innovation has always been perceived as the realm of the production industry for its survival in a competitive market (Marleba, 2006), there is a need for its robust articulation as a requirement in IHL. Countries in Africa have lagged behind in originating their own intellectual property, but they have the advantage of leveraging existing knowledge as a platform to work from and advance their own inventions into innovations to solve problems peculiar to Africa. Africa is both a huge resource base and a market with sufficient opportunity for exploitation by its citizens.

In order to successfully implement innovation in African IHLs that have sorely lagged behind, it is important to have a succinct and crystal-clear understanding by the faculty of how research, innovation and enterprise are linked. The success of building the university of future also lies in fostering a win–win collaboration between industry, government, non-governmental organisations and development agencies, allied with a diligent faculty capable of aligning their activities with those of other players in the ecosystem.

It is when this new knowledge is taken, used and applied for problem-solving that an invention becomes an innovation. Additionally, when old knowledge is legitimately applied as new knowledge in a different context, it constitutes another form of innovation, as the use of this knowledge leads to solutions to problems.

References

Abramo, G., & Costa, F. D. (2011). University-industry research collaboration: A model to assess. *Higher Education, 62*(2011), 163–181. https://doi.org/10.1007/s10734-010-9372-0

Afshin, A., Sur, P. J., Fay, K. A., Cornaby, L., Ferrara, G., Salama, J. S., Mullany, E. C., Abate, K. H., Abbafati, C., Abebe, Z., Afarideh, M., Aggarwal, A., Agrawal, S., Akinyemiju, T., Alahdab, F., Bacha, U., Bachman, V. F., Badali, H., Badawi, A., Bensenor, I. M., et al. (2019). Health effects of dietary risks in 195 countries, 1990–2017: A systematic analysis for the Global Burden of Disease Study 2017. *The Lancet, 393*(10184), 1958–1972. https://doi.org/10.1016/S0140-6736(19)30041-8

Agwu, A. E., Dimelu, M. U., & Madukwe, M. C. (2008). Innovation system approach to agricultural development: Policy implications for agricultural extension delivery in Nigeria. *African Journal of Biotechnology, 7*(11), 1604–1611. https://doi.org/10.5897/AJB08.289

Albulescu, I., & Albulescu, M. (2014). The University in the community. The university's contribution to local and regional development by providing educational services for adults. *Procedia - Social and Behavioral Sciences, 142*, 5–11. https://doi.org/10.1016/j.sbspro.2014.07.578

Allison, C. B., Breimyer, H. F., Lambert, W. N., Leuthold, F. O., & Martin, J. A. (1976). *The agrarian tradition in American Society.*

Andoh, H. (2017). Research: The "Lost Mission" of African Universities. *International Higher Education, 91*, 20–22. https://doi.org/10.6017/ihe.2017.91.10133

Ankrah, S., & Al-Tabbaa, O. (2018). Universities-industry collaboration: A systematic review. *Scandinavian Journal of Management, 31*(3), 387–408. https://doi.org/10.1016/j.scaman.2015.02.003

Aristovnik, A., Keržič, D., Ravšelj, D., Tomaževič, N., & Umek, L. (2020). Impacts of the COVID-19 pandemic on life of higher education students: A global perspective. *Sustainability (Switzerland), 12*(20), 1–34. https://doi.org/10.3390/su12208438

Armbruster, C. (2008). Research universities: Autonomy and self-reliance after the entrepreneurial university. *Policy Futures in Education, 6*(4), 372–389. https://doi.org/10.2304/pfie.2008.6.4.372

Awasthy, R., Flint, S., Sankarnarayana, R., & Jones, R. L. (2020). A framework to improve university–industry collaboration. *Journal of Industry-University Collaboration, 2*(1), 49–62. https://doi.org/10.1108/jiuc-09-2019-0016

Banerjee, C. (2014). *The human facto: The fundamental driver of innovation.*
Bank, W. (2012). *Agricultural innovation systems.* https://doi.org/10.1596/978-0-8213-8684-2
Bar-On, Y. M., Phillips, R., & Milo, R. (2018). The biomass distribution on Earth. *Proceedings of the National Academy of Sciences of the United States of America, 115*(25), 6506–6511. https://doi.org/10.1073/pnas.1711842115
Barasa, L., Knoben, J., Vermeulen, P., Kimuyu, P., & Kinyanjui, B. (2017). Institutions, resources and innovation in East Africa: A firm level approach. *Research Policy, 46*(1), 280–291. https://doi.org/10.1016/j.respol.2016.11.008
Bartels, F. L., Koria, R., & Vitali, E. (2016). Barriers to innovation: The case of Ghana and implications for developing countries. *Triple Helix, 3*(1). https://doi.org/10.1186/s40604-016-0040-y
Bas, T. G., & Kunc, M. H. (2009). National systems of innovations and natural resources clusters: Evidence from copper mining industry patents. *European Planning Studies, 17*(12), 1861–1879. https://doi.org/10.1080/09654310903322363
Berger, M., Murugi, J., Buch, E., Usselmuiden, C., Kennedy, A., Moran, M., Guzman, J., Devlin, M., & Kubata, B. (2009). *Strengthening pharmaceutical innovation in Africa designing strategies for national pharmaceutical innovation.* Council on Health Research for Development (COHRED); New Partnership for Africa's Development (NEPAD).
Bhayani, A. (2015). Building entrepreneurial universities in a specific culture-barriers and opportunities. *International Journal of Nonprofit and Voluntary Sector Marketing, 20*(4), 312–330. https://doi.org/10.1002/nvsm.1533
Bigirimana, S. (2021). Embedding intellectual property in university education: Strides towards education 5.0. *ACADEMIA Letters,* 0–5. https://doi.org/10.20935/AL3035
Brennan, J., Broek, S., Durazzi, N., Kamphuis, B., Ranga, M., & Ryan, S. (2014). Study on innovation in higher education: Final report. In *European Commission Directorate for Education and Training Study on Innovation in Higher Education.*
Cai, Y., Ma, J., & Chen, Q. (2020). Higher education in innovation ecosystems. *Sustainability (Switzerland), 12*(11), 1–12. https://doi.org/10.3390/su12114376
Celeste, R. F., Griswold, A., & Straf, M. L. (2014). Understanding the pathways from research to innovation. In *Furthering America's Research Enterprise* (pp. 1–200). National Academy of Sciences. https://doi.org/10.17226/18804
Chetsanga, C. J. (2003). *An analysis of the cause and effect of the brain drain in Zimbabwe.*
Chinyoka, A., & Mutambara, E. (2020). The challenges of revenue generation in state universities: The case of Zimbabwe. *Cogent Social Sciences, 6*(1). https://doi.org/10.1080/23311886.2020.1748477
Chitate, H. (2018). Science, technology, engineering and mathematics (STEM): A case study of Zimbabwe's educational approach to industrialisation. *World Journal of Education, 6*(5), 27–35. https://doi.org/10.5430/wje.v6n5p27
Consoli, D., & Mina, A. (2009). An evolutionary perspective on health innovation systems. *Journal of Evolutionary Economics, 19*(2), 297–319. https://doi.org/10.1007/s00191-008-0127-3
Daniels, R. J. (2020). *Universities' vital role in the pandemic response.* John Hopkins University.
Dessler, A. J. (1969). The role of basic research in universities. *Eos, Transactions American Geophysical Union, 50*(9), 508–511. https://doi.org/10.1029/EO050i009p00508
Diop, M. (2017). *Innovation in Africa.* World Bank.
Dubiński, J. (2013). Sustainable development of mining mineral resources. *Journal of Sustainable Mining, 12*(1), 1–6. https://doi.org/10.7424/jsm130102
Dziwa, D. D., & Postma, L. (2020). Building creative capacities through art teacher education in Zimbabwe. *Bulgarian Comparative Education Society (BCES), 18,* 277–283.
Fehner, B. T. R., & Gosling, F. G. (2012). *The Manhattan Project—ThingLink. April.*
Gibb, A., Haskins, G., & Robertson, I. (2009). *Leading the entrepreneurial meeting the entrepreneurial development needs of higher education institutions* (pp. 1–44). National Council for Graduate Entrepreneurship (NCGE) & Said Business School, University of Oxford.

Ginzberg, E., & Vojta, G. J. (1981). The service sector of the U.S. Economy. *Scientific American, 244*(3), 48–55. https://doi.org/10.1038/scientificamerican0381-48
Godin, B. (2006). The linear model of innovation. *Science, Technology, & Human Values, 3*(6), 639–667. https://doi.org/10.1177/0162243906291865
Godin, B. (2007). *National innovation system: The system approach in historical perspective Benoît Godin 385 rue Sherbrooke Est Montreal, Quebec Canada H2X 1E3 Project on the History and sociology of STI statistics working paper No. 36 Previous Papers in the Series: 36*, 1–37.
Godin, B. (2008). *Innovation: The history of a Category.*
Godin, B. (2010). *National innovation system: A note on the origins of a concept Benoît Godin 385 rue Sherbrooke Est Montreal, Quebec Canada H2X 1E3 Project on the Intellectual history of innovation working papers.*
Goonetilleke, A., & Vithanage, M. (2017). Water resources management: Innovation and challenges in a changing world. *Water (Switzerland), 9*(4). https://doi.org/10.3390/w9040281
Griliches, Z. (1990). Patent statistics as economic indicators: A survey. *Journal of Economic Literature, 28*(4), 1661–1707.
Gunasekara, C. (2006). Reframing the role of Universities in the development of regional innovation systems. *Journal of Technology Transfer, 31*(1), 101–113. https://doi.org/10.1007/s10961-005-5016-4
Guvhu, R., & Museva, L. (2020). Exploring management styles of university department chairpersons in improving the quality of education in Zimbabwe. In *International conference on business management, innovation, and sustainability, September 2015.*
Harthy, H. S., et al. (2014). *The entrepreneurial university and the entrepreneurial environment: Organizational analysis and policy considerations A thesis submitted to The University of Manchester for the degree of Doctor of Philosophy in the Faculty of Humanities.* University of Manchester.
International Telecommunication Union. (2015). The Internet of Things: Data for development. In *Measuring the information society report 2015.*
James, W. P. T. (2008). WHO recognition of the global obesity epidemic. *International Journal of Obesity, 32*, S120–S126. https://doi.org/10.1038/ijo.2008.247
Jengeta, M. (2020). *Effectiveness of entrepreneurship education in a turbulent economy: The perceptions of entrepreneurship graduates in Zimbabwe.* University of Kwazulu Natal.
Juma, C. (2011). Growing the economy. In *The New Harvest* (p. 4). Oxford University Press.
Kaviya, M., Nikisi, D., Manuere, F., & Sciences, B. (2022). *Involvement of academic staff in developing higher education curriculum for human capital training in Zimbabwe* (pp. 228–242).
Keche, K. (2021). Relevancy of new higher education approaches in 'Second Republic Zimbabwe'. In L. Waller & S. Waller (Eds.), *Higher education—New approaches to accreditation, digitalization, and globalization in the age of COVID.* IntechOpen.
Kim, J. H., & Scialli, A. R. (2011). Thalidomide: The tragedy of birth defects and the effective treatment of disease. *Toxicological Sciences, 122*(1), 1–6. https://doi.org/10.1093/toxsci/kfr088
Kinkel, S., Lay, G., & Wengel, J. (2005). *Innovation: More than research and development growth opportunities on different innovation paths.*
Léger, A., & Swaminathan, S. (2007). *Discussion papers innovation theories: Relevance and implications for developing country innovation.*
Mačiulis, A., Vasiliauskas, A. V., & Jakubauskas, G. (2009). The impact of transport on the competitiveness of national economy. *Transport, 24*(2), 93–99. https://doi.org/10.3846/1648-4142.2009.24.93-99
Mahdjoubi, D. (1997). *The non-linear model of technological innovation.*
Mamina, M. T., & Maganga, R. (2019). A review of engineering education at the University of Zimbabwe. *African Journal of Engineering Research, 7*, 112–128.
Manley, K. (2002). The systems approach to innovation studies. *Australasian Journal of Information Systems, 9*(2). https://doi.org/10.3127/ajis.v9i2.196
Maré, D. C., Fabling, R., & Stillman, S. (2011). *Immigration and innovation Author contact details. May.*

Marleba, F. (2006). *Innovation, industrial dynamics and industry evolution: Progress and the research agendas Franco Malerba OFCE | « Revue de l ' OFCE » 2006/5 no 97 bis | pages 21 à 46 ISSN 1265-9576 ISBN 2-7246-3066-1 Article disponible en ligne à l' adresse*

Martinuzzi, A., Blok, V., Brem, A., Stahl, B., & Schönherr, N. (2018). Responsible research and innovation in industry-challenges, insights and perspectives. *Sustainability (Switzerland), 10*(3), 1–9. https://doi.org/10.3390/su10030702

Miguel Benavente, J. (2006). The role of research and innovation in promoting productivity in Chile. *Economics of Innovation and New Technology, 15*(4–5), 301–315. https://doi.org/10.1080/10438590500512794

Ministry of Higher and Education Innovation Science and Technology Developments. (2018). *Education 5.0—Towards problem-solving and value creation*. Ministry of Higher and Education Innovation Science and Technology Developments.

Ministry of Higher and Tertiary Education Innovation Science and Technology Development. (2019). *Education 5.0 Doctrine for the modernisation and industrialisation of Zimbabwe through education, science and technology development to archive Vision 2030: Industrialisation.*

Molella, A., & Kargon, R. (2004). The city as communications net: Norbert Wiener, the atomic bomb, and urban dispersal. *Technology and Culture, 45*(4), 764–777.

Mungazi, D. A. (1985). Educational innovation in Zimbabwe: Possibilities and problems. *Journal of Negro Education, 54*(2), 196–212.

Murairwa, S. (2021). A sustainable Zimbabwe University—Industry collaboration framework. *International Journal of Research and Innovation in Social Science, 7*(4), 112–128.

Murgor, T. K. (2015). Challenges facing adoption of information communication technology in African universities. *Journal of Education and Practice, 6*(25), 62–68.

Musindo, B., & Mudavanhu, V. (2006). Entrepreneurship education at Universities.pdf. *Southern Africa Journal of Education, Science and Technology, 1*(1). https://doi.org/10.4314/sajest.v1i1.39797

Muyambo, T., & Ndamba, G. T. (2021). The status of indigenous languages in higher education: A case study of one university in Zimbabwe. *African Perspectives of Research in Teaching & Learning, 5*(2).

Muzira, D. R., & Bondai, B. M. (2020). Perception of educators towards the adoption of Education 5.0: A case of a state university in Zimbabwe. *East African Journal of Education and Social Sciences, 1*(2), 43–53.

Muzira, D. R., & Muzira, R. (2020). An assessment of educators' level of concern on the adoption of Education 5.0: A case of one university in Zimbabwe. *Current Journal of Applied Science and Technology*. https://doi.org/10.9734/cjast/2020/v39i1730749

Naidoo, N. (2011). What is research? A conceptual understanding. *African Journal of Emergency Medicine, 1*(1), 47–48. https://doi.org/10.1016/j.afjem.2011.05.011

Nicolaides, A. (2014). Research and innovation—The drivers of economic development. *African Journal of Hospitality, Tourism and Leisure, 3*(2), 1–16.

Poh, L. C. (2016). *Chapter 10: From research to innovation to enterprise: The case of Singapore 133 By Lim Chuan Poh, Agency for Science, Technology and Research (A*STAR).*

Pticek, M., Podobnik, V., & Jezic, G. (2016). Beyond the Internet of Things: The social networking of machines. *International Journal of Distributed Sensor Networks, 2016*. https://doi.org/10.1155/2016/8178417

Pugh, R., Lamine, W., Jack, S., & Hamilton, E. (2018). The entrepreneurial university and the region: What role for entrepreneurship departments? *European Planning Studies, 26*(9), 1835–1855. https://doi.org/10.1080/09654313.2018.1447551

QUB. (2022). *Research output*. QUB.

QUBIS. (2015a). *About us*. QUBIS.

QUBIS. (2015b). *From research to wealth*. QUBIS.

QUBIS Ltd Belfast. (2016). *Peter Gregson interview*. QUBIS Ltd Belfast.

Republic of Zimbabwe. (2021). *"Towards a prosperous & empowered upper middle income society by 2030" NATIONAL.*

Research Council of Zimbabwe. (2017). *11th Zimbabwe International research symposium book of papers presented*. Research Council of Zimbabwe.
Robinson, K. B. (2020). *Perspectives of highly skilled migrants on return migration: A qualitative case study of Zimbabwean lecturers in the Western Cape of South Africa*. University of the Western Cape.
Samadi, H., Researchers, Y., Club, E., Branch, Q., Azad, I., & Samadi, H. (2018). *Entrepreneurial university, a necessity for knowledge-based economy ; evaluation and explanation of entrepreneurial capacity of University of Mazandaran, 8*(3), 53–66.
Sawyerr, A. (2017). African universities and the challenge of research capacity development. *Journal of Higher Education in Africa, 2*(1), 213–242. Revue de l' enseignement supérieur en Stable. http://www.jstor.org/stable/24486132
Schmidt, L. B. (2021). *Whither agrarian economy in the United States* (Vol. 15, No. 2, pp. 196–205). Oxford University Press. Stable URL: https://www.jstor.org/stable/2570956
Scott, G., Coates, H., & Anderson, M. (2008). *Learning leaders in times of change: Academic leadership capabilities for Australian higher education.*
Scott, R., & Vincent-Lancrin, S. (2014). Chapter 3: Educating innovators and entrepreneurs By Richard Scott and Stéphan Vincent-Lancrin, OECD Directorate for Education and Skills. In S. Dutta, B. Lanvin, & S. Wunsch-Vincent (Eds.), *The global innovation index 2014-The human factor in innovation.*
Sivaram, V., Dabiri, J. O., & Hart, D. M. (2018). The need for continued innovation in solar, wind, and energy storage. *Joule, 2*(9), 1639–1642. https://doi.org/10.1016/j.joule.2018.07.025
Stock, R. M. (2014). Impacts of hedonic and utilitarian user motives on the innovativeness of user-developed solutions. *Journal of Product Innovation Management, 32*(3), 389–403.
Taylor, S. P. (2017). What is innovation? A study of the definitions, academic models and applicability of innovation to an example of social housing in England. *Open Journal of Social Sciences, 05*(11), 128–146. https://doi.org/10.4236/jss.2017.511010
Thapa, R. K., Iakovleva, T., & Foss, L. (2019). Responsible research and innovation: A systematic review of the literature and its applications to regional studies. *European Planning Studies, 27*(12), 2470–2490. https://doi.org/10.1080/09654313.2019.1625871
Universities UK. (2015). *The economic impact of Queen's University Belfast on the Northern Ireland economy.*
Uyarra, E. (2008). *The impact of universities on regional innovation: A critique and policy implications, Manchester Business School Working Paper* (564).
Vidican, G. (2015). The emergence of a solar energy innovation system in Morocco: A governance perspective. *Innovation and Development, 5*(2), 225–240. https://doi.org/10.1080/2157930X.2015.1049852
White, E. G. (1888). Chapter 15—The Bible and the French Revolution. In Ellen G White Estate (Ed.), *The great controversy* (p. 279). Ellen G. White Estate, Inc.
White, E. G. (1905). Chapter 12—Help for the unemployed and the homeless. In Ellen G White Estate (Ed.), *The ministry of healing* (p. 187). Ellen G. White Estate, Inc.
WIPO. (2014). *Understanding industrial property rights.*
WIPO. (2021). *Global Innovation Index 2021-Zimbabwe.*
Xiangjiang, Q., Peng, J., & Kelly, J. (2013). Open innovation: The view of an ICT leader in distributed global innovation. In *The Global Innovation Index 2013*. World Intellectual Property Organisation.
Zvobgo, R. J. (2019). Reforms in higher education in Africa: Revitalising Africa's higher education systems through research. *Journal of New Vision in Educational Research, 1.*

Chapter 7
Exploring Methods of Anaemia Identification Using Haemoglobin Levels for ASEAN Countries

Htet Ne Oo, Heena Wadhwa, Mandeep Kaur, Saurav Kumar, and Neha

Abstract Haemoglobin plays a crucial role in the human body in keeping someone fit and healthy. In developing countries, a huge number of people lack basic health facilities, and the number of accessible doctors is less. Hence, there is a need for low-cost or free healthcare devices. The non-invasive haemoglobin measurement device is one example. It enables measurement without invasion, eliminating the chances of infection. Apart from this, it can be easily used by semi-skilled volunteers. The GPS-enabled device will help tag the location of individuals with their haemoglobin values. The Government can obtain real-time haemoglobin data using this device. The innovation provides a technological solution to a fundamental problem facing the population. It will reach the population via Government health services. The biggest contribution of this innovation is the creation of a Haemoglobin map for the surveyed area. This will be the first of its kind in the country, to the best of our knowledge, with current technology. This innovation will bring a paradigm shift in the policy ecosystem. Government policy can be effectively implemented on the ground, and its impact can be monitored through the real-time map. Haemoglobin measurement can be done without invasive techniques, which can cause infection in infants and pregnant women. This innovation will be a small step towards Digital and Healthy ASEAN.

Keywords Haemoglobin · Artificial intelligence · GPS · Health monitoring · Healthy ecosystem

---

H. N. Oo (✉) · H. Wadhwa · M. Kaur
Department of Computer Science and Engineering, Chitkara University Institute of Engineering and Technology, Chitkara University, Punjab, India
e-mail: htetneoo@gmail.com; htet.neo@chitkara.edu.in

S. Kumar · Neha
Council of Scientific and Industrial Research, Central Scientific Instruments Organization, Sector 30 C, Chandigarh, India
e-mail: sauravpandey.csio@csir.res.in

S. Kumar
Academy of Scientific and Innovative Research, Ghaziabad, Uttar Pradesh, India

X. Poshiwa et al. (eds.), *Technological Innovations for Sustainable Industrialization*,
https://doi.org/10.1007/978-981-96-7995-9_7

Introduction

Haemoglobin is one of the most important factors for human health. To maintain good health, people need to regularly check their blood pressure, haemoglobin level, sugar level, and other important factors. Haemoglobin can be measured with laboratory analysis, and it is a common indicator of anaemia. Many genetic haematological diseases are very common worldwide. There is a wide occurrence of genetic blood diseases in Southeast Asian countries. To check haemoglobin levels, an invasive method involving blood draws is typically used. The diagnosis of anaemia typically involves measuring the haemoglobin level in the blood. If the haemoglobin level is below the normal range, additional tests may be performed to determine the underlying cause of the anaemia. Haemoglobin measurement is a critical tool in the diagnosis, monitoring, and management of various medical conditions. It can be used for diagnosing of anaemia; to monitor response to treatment; in preoperative evaluation, screening for blood disorders and monitoring chronic diseases. It provides important information about the oxygen-carrying capacity of the blood and can guide appropriate treatment decisions. There are some common methods used to measure haemoglobin levels in the blood. The choice of method depends on various factors, including the purpose of the test, the available resources, and the clinical situation. Some well-known methods are complete blood count (CBC), Haemoglobin A1c (HbA1c), Haemoglobin Electrophoresis, Spectrophotometry, Point-of-care Testing (POCT) and the Cyanmethemoglobin method. All of these methods are invasive and require a blood sample. Apart from these invasive methods, there are also non-invasive methods. However, the results from invasive methods are more accurate than those from non-invasive methods. More studies and research are required on non-invasive methods. It is important for everyone to stay fit. And to monitor health and fitness, haemoglobin plays a crucial role. Based on population health studies in different countries, it is found that anaemia is one of the serious concerns in many countries. Therefore, this paper seeks to analyse haemoglobin value patterns across ASEAN countries to identify anaemia prevalence, explore and compare different diagnostic methods for early anaemia detection, and propose region-specific strategies for enhancing screening and healthcare interventions. Moreover, this paper aims to find out the limitations of current haemoglobin colour scale and suggest the importance of non-invasive methods of investigating haemoglobin values. Furthermore, the study provides a future direction for developing a comprehensive haemoglobin map for ASEAN countries to support evidence-based and targeted public health planning.

Literature Review

ASEAN Countries Under Anaemia Threat

In many Association of Southeast Asian Nations (ASEAN) countries, one of the most significant health concerns is Anaemia. It occurs especially in women and children. According to a World Health Organisation (WHO) report from 2019, 29.9% of women aged 15–49 years suffered from Anaemia. The same report indicated that a significant percentage of children aged 6–59 months, also suffered from Anaemia. The prevalence of Anaemia varies across specific ASEAN countries. According to the same report by WHO, the prevalence of Anaemia among women of reproductive ranges from 15% in Thailand to 47% in Laos. Among pregnant women, the prevalence ranges from 17% in Malaysia to 62% in Cambodia. The prevalence of Anaemia in children under five years of age ranges from 8% in Thailand to 58% in Timor-Leste. Therefore, it can be said that Anaemia is a crucial health issue in many ASEAN countries. Figure 7.1 shows the occurrence of Anaemia in different countries. Support for nutrition and access to iron supplements are essential in those regions, especially for women and children, to prevent Anaemia. This is a significant health concern as Anaemia can lead to a range of health problems, including fatigue, weakness and impaired cognitive function (Van den Broek et al., 1999).There are several initiatives in place to address Anaemia in ASEAN countries, including efforts to improve nutrition and increase access to iron supplements. However, more work is needed to raise awareness about the issue and ensure that all individuals have access to the resources they need to prevent and treat anaemia (Sunuwar et al., 2020).

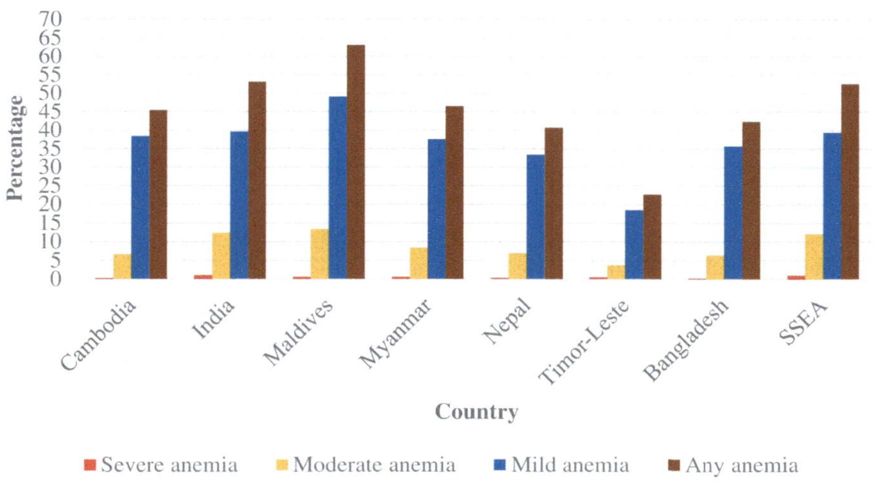

Fig. 7.1 Prevalence of any form of anaemia among women of reproductive age by country (Sunuwar et al., 2020)

There are different methods to measure haemoglobin levels in the blood. The most common methods are invasive methods, which require drawing blood. Some other methods are non-invasive methods, which do not require drawing blood.

Common Method of Estimation of Anaemia in ASEAN Countries

To check the possibility of anaemia, the most commonly used method is to measure the haemoglobin (Hb) value in the blood. Haemoglobin is a protein in the red blood cells that carries oxygen throughout the body. A low level of haemoglobin indicates that there are not enough red blood cells in the blood or that the existing cells are not working properly, which can cause anaemia. To measure haemoglobin level, a simple blood test called a complete blood count (CBC) is used. The number and size of red blood cells and the amount of haemoglobin are measured in the test. If the haemoglobin level is below the normal range, it is called anaemia. Normal results for a haemoglobin test for adults vary, but generally are:

- Men: 13.8 to 17.2 g per decilitre (g/dL) or 138 to 172 g per litre (g/L)
- Women: 12.1 to 15.1 g/dL or 121 to 151 g/L
- Newborn: 14 to 24 g/dL or 140 to 240 g/L
- Infant: 9.5to 13 g/dL or 95 to 130 g/L.

Another useful method for diagnosing anaemia is measuring the haematocrit level, which is the percentage of red blood cells in the blood, and counting the number of red blood cells in a blood sample.

The measurement of haemoglobin (Hb) level in the blood is the most commonly used method to determine the possibility of anaemia. Haemoglobin is a protein present in red blood cells that transports oxygen throughout the body. A low haemoglobin level indicates a deficiency of red blood cells or malfunctioning of existing cells, which can cause anaemia. A complete blood count (CBC) is a simple blood test used to measure haemoglobin level and the number and size of red blood cells. Anaemia is diagnosed when the haemoglobin level falls below the normal range. Normal haemoglobin levels for adults vary by gender and age. Measuring the haematocrit level, which is the percentage of red blood cells in the blood, and counting the number of red blood cells in a blood sample are other useful methods for diagnosing anaemia. Anaemia is diagnosed when the haemoglobin level falls below the normal range (Horowitz et al., 2013). The normal results of a haemoglobin test for different age groups and genders are as follows (Table 7.1).

Table 7.1 Range of haemoglobin in different population groups

Population	Haemoglobin levels (g/dL)
Male	13.8 to 17.2
Female	12.1 to 15.1
Newborn	14 to 24
Infant	9.5 to 13

Historical Data of Anaemia

The systematic study of the global burden of anaemia reveals compelling insights into its pervasive impact across different demographics and regions. Despite notable declines in prevalence from 1990 to 2010, anaemia remains a significant health concern, ranking among the top contributors to years lived with disability (YLD). In 2010 alone, anaemia accounted for 68.3 million YLD worldwide, surpassing the burden of major depression, chronic respiratory diseases and injuries. The prevalence of anaemia varied across age groups, with children under 5 years old exhibiting particularly high rates, suggesting a critical need for targeted interventions in this vulnerable population. Moreover, gender disparities in anaemia prevalence have widened over time, with females experiencing higher rates across all age groups and regions. Factors contributing to anaemia include poor nutrition, parasitic infections and limited access to healthcare services and interventions. Efforts to address anaemia should prioritise optimising nutrition, implementing iron supplementation programmes and targeting underlying causes, especially in high-risk populations. Additionally, the study underscores the importance of targeted surveillance and intervention strategies to effectively mitigate the burden of anaemia, particularly in regions with the highest prevalence rates. By providing comprehensive estimates and insights, this study serves as a valuable resource for healthcare specialists, policymakers and donors seeking to develop evidence-based strategies for anaemia prevention and control.

Anaemia poses a significant health challenge in ASEAN countries, especially among women of reproductive age (15–49 years) and children under 5 years old. According to data compiled by the World Health Organisation (WHO) in 2016, the prevalence of anaemia varied across the region (Table 7.2).

Despite some progress, anaemia rates persist at high levels in several ASEAN countries, particularly affecting women of reproductive age and young children. Contributing factors include poor nutrition, parasitic infections and limited access to health services such as iron supplementation and deworming interventions. Ongoing efforts to combat anaemia in ASEAN countries involve initiatives to enhance nutrition, strengthen health services, promote food fortification and increase access to interventions for anaemia prevention and treatment.

As per the World Bank, the prevalence of anaemia among women aged 15–49 in Myanmar was 42.7% in 1995, which declined to 33.3% in 2016. Similarly, among children under 5 years of age, anaemia prevalence was 74.2% in 1995 and

Table 7.2 Prevalence among women of reproductive age (15–49 years) and children under 5 years old

Country	Prevalence among women (%)	Prevalence among children (%)
Cambodia	32.5	43.1
Indonesia	28.1	34.8
Lao PDR	42.9	47.5
Malaysia	17.7	7.7
Myanmar	33.3	44.5
Philippines	30.2	40.1
Thailand	21.9	28.5
Vietnam	23.1	23.3

significantly reduced to 44.5% by 2016. Findings from the Global Nutrition Report 2020 indicate that the prevalence of anaemia among women of reproductive age in Myanmar was 31.6% in 2017, while the prevalence among children under 5 years of age was 43.4% in the same year (Sunuwar et al., 2020).

Historical Data of Anaemia in ASEAN Countries

According to data from the World Health Organisation (WHO), the prevalence of Anaemia among women of reproductive age (15–49 years) in ASEAN countries varied widely in 2016. The prevalence of Anaemia was:

- Cambodia: 32.5%
- Indonesia: 28.1%
- Lao PDR: 42.9%
- Malaysia: 17.7%
- Myanmar: 33.3%
- Philippines: 30.2%
- Thailand: 21.9%
- Vietnam: 23.1%.

Among children under 5 years of age, the prevalence of anaemia in ASEAN countries was also high, with the following rates reported by the WHO in 2016:

- Cambodia: 43.1%
- Indonesia: 34.8%
- Lao PDR: 47.5%
- Malaysia: 7.7%
- Myanmar: 44.5%
- Philippines: 40.1%
- Thailand: 28.5%

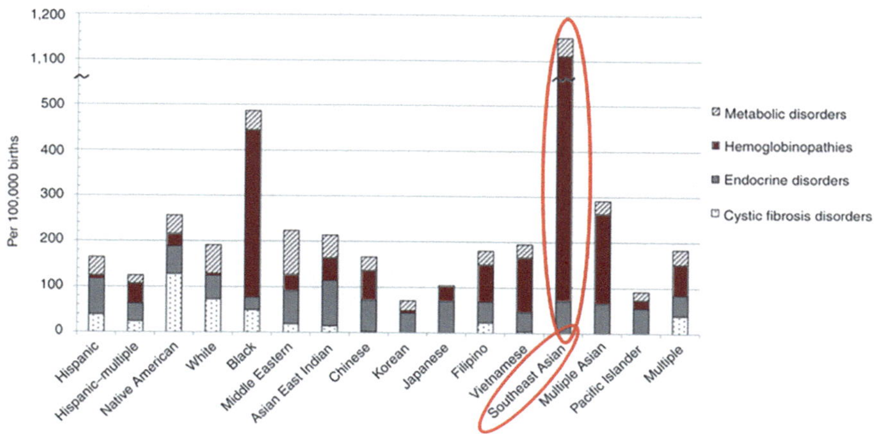

Fig. 7.2 Haemoglobin issues in different regions

- Vietnam: 23.3%.

While anaemia rates have decreased in some ASEAN countries over time, the prevalence of anaemia remains high in many countries, particularly among women of reproductive age and children, as shown in Figs. 7.2 and 7.3. Factors contributing to anaemia in ASEAN countries include poor nutrition, parasitic infections and limited access to health services and interventions such as iron supplementation and deworming. Efforts to address anaemia in ASEAN countries include improving nutrition and health services, promoting food fortification and increasing access to interventions for anaemia prevention and treatment.

According to the World Bank, the prevalence of anaemia among women of reproductive age (15–49 years) in Myanmar was 42.7% in 1995, which decreased to 33.3% in 2016. Among children under 5 years of age, the prevalence of anaemia was 74.2% in 1995, which decreased to 44.5% in 2016. The Global Nutrition Report 2020 also reported that the prevalence of anaemia among women of reproductive age in Myanmar was 31.6% in 2017, while the prevalence of anaemia among children under 5 years of age was 43.4% in 2017 (Sunuwar et al., 2020) (Table 7.3).

Different Methods of Anaemia Detection Based on Haemoglobin Levels

There are various methods that can be used for anaemia detection from haemoglobin levels. The two broad categories are invasive methods and non-invasive methods. Some methods are based on the development of colour. Some methods use reagents to mix with blood to produce particular results.

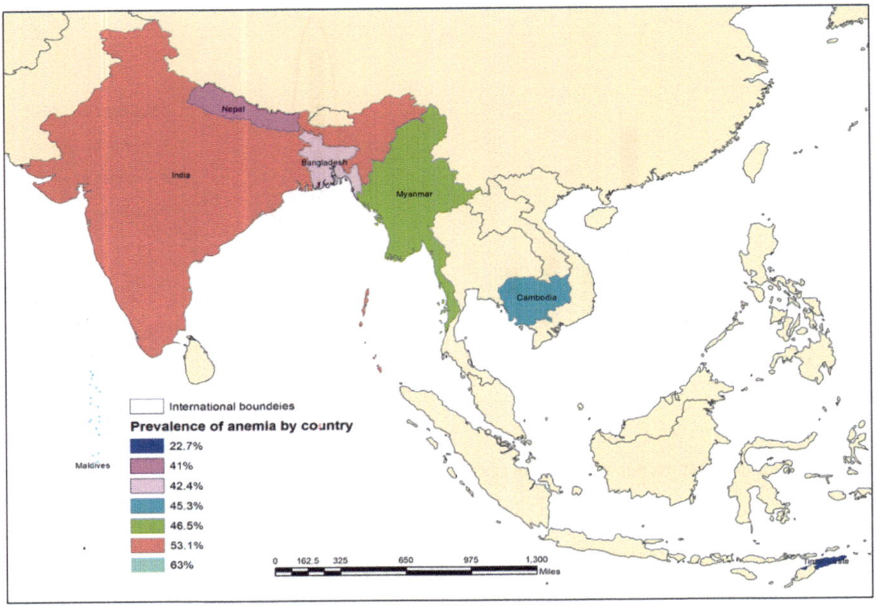

Fig. 7.3 Map showing the prevalence of anaemia among women of reproductive age in seven selected South and Southeast Asian countries

Some low-cost methods that are used in ASEAN countries for estimating haemoglobin value are: Some low-cost methods that are used in ASEAN countries for estimating haemoglobin value are:

- Haemoglobin colour scale: In this method, a colour scale is used to estimate haemoglobin levels. A reagent solution is used to mix with a drop of blood and the result is compared with a colour chart to determine the haemoglobin level.
- Copper sulphate method: In this traditional method, which has been used for many years to estimate haemoglobin levels, a drop of blood is mixed with a copper sulphate solution, and depending on the haemoglobin level, the colour of the solution changes.
- Sahli's method: This is a simple and less expensive method that uses a glass tube with markings of Hb on one side. Blood is drawn into the tube and mixed with a reagent and distilled water, and the level of haemoglobin is read off a graduated scale on the tube.
- Haemoglobinometer: A photometric method is used to measure haemoglobin level in a drop of blood in a portable device. Minimal training is required, and it is simple to use.
- Cyanmethaemoglobin method: This method is internationally recommended for measuring haemoglobin. A spectrophotometer and reagents are required to measure haemoglobin. The accuracy of these methods may be less than that of

Table 7.3 Effect of anaemia on women and children in Southeast Asian countries

Country	Demographic group	Prevalence of anaemia	Causes of anaemia	Health outcomes	References	Interpretation
India	Children	60.13%	Poor nutrition intake, lower maternal haemoglobin level, infections with malaria parasite, hookworm	Impaired cognitive and psychomotor development	Onyeneho et al. (2019)	Anaemia in Indian children is primarily caused by factors like inadequate nutrition and maternal health issues, which can result in developmental impairments
India	Women	50%	Nutrient deficiency due to inadequate diets and absorption, infections such as malaria, chronic diseases	Loss of normal skin colour, impaired fetal development during pregnancy, reduced physical capacity	Kaur (2014)	Anaemia among Indian women is attributed to factors like nutrient deficiency and infections, which leadi to adverse effects on health and pregnancy outcomes
Bangladesh	Children	53% (male), 51% (female)	Poor nutritional intake of iron, low iron bioavailability, maternal anaemia	Negative impacts on central nervous system, immune system, cognitive and physical development	Ali et al. (2021)	Anaemia prevalence among Bangladeshi children is influenced by factors like iron intake and maternal health, which can affect neurodevelopment and growth
Bangladesh, Nepal, Maldives	Women of reproductive age	41.8% (Bangladesh), 58.5% (Maldives), 40.6% (Nepal)	Nutrition deficiency, chronic infections, genetic disorders, underweight, pregnancy status	Increased risk of miscarriages, premature birth, low birth weight, impaired learning	Rahman et al. (2021)	Anaemia prevalence among women in these countries is linked to various factors, which impact maternal and child health outcomes

(continued)

Table 7.3 (continued)

Country	Demographic group	Prevalence of anaemia	Causes of anaemia	Health outcomes	References	Interpretation
Indonesia	Women of reproductive age and children under 5	28.1% (women), 34.8% (children)	Iron deficiency during pregnancy, diabetes, maternal smoking, preterm birth, low birth weight, multiple pregnancies	Reduced physical and mental capacity, increased health risks	Sari et al. (2022) and Margawati et al. (2023)	Anaemia in Indonesia is associated with factors like pregnancy complications and lifestyle habits, which can lead to health complications
Cambodia	Women of reproductive age and children under 5	32.5% (women), 43.1% (children)	Folic acid deficiencies, haemoglobinopathies (genetic disorder)	Negative impacts on physical health, increased mortality risk in severely anaemic children	Wieringa et al. (2016)	Anaemia prevalence in Cambodia is influenced by genetic disorders and nutritional deficiencies, which impact mortality and health outcomes
Myanmar	Women of reproductive age and children under 5	33.3% (women), 44.5% (children)	Inadequate nutrition, infectious diseases (malaria, hookworm), socioeconomic factors, genetic disorders	Increased infection risk, compromised immune system	Win and Ko (2018)	Anaemia in Myanmar is attributed to factors like nutrition and infectious diseases, which impact immune function and overall health
Malaysia	Women of reproductive age and children under 5	17.7% (women), 7.7% (children)	Socioeconomic factors, maternal age extremes, pregnancy status	Impaired development, preterm delivery	Nik Rosmawati et al. (2012)	Anaemia in Malaysia is associated with socioeconomic factors and pregnancy status, which affects maternal and child health outcomes
Philippines	Women of reproductive age and children under 5	30.2% (women), 40.1% (children)	Abnormal haemoglobin levels	Physical health effects, increased infection susceptibility	Goyena et al. (2020)	Anaemia in the Philippines is linked to abnormal haemoglobin levels, which impact physical health and infection resistance

(continued)

Table 7.3 (continued)

Country	Demographic group	Prevalence of anaemia	Causes of anaemia	Health outcomes	References	Interpretation
Lao PDR	Women of reproductive age and children under 5	42.9% (women), 47.5% (children)	Unsafe water, poor sanitation, pregnancy status	Adverse pregnancy outcomes, neonatal health risks	Keokenchanh et al. (2021)	Anaemia prevalence in Laos is influenced by factors like water quality and pregnancy, which affect maternal and child health
Thailand	Women of reproductive age and children under 5	21.9% (women), 28.5% (children)	Thalassemia	Weakness, shortness of breath, fatigue	Jamnok et al. (2020)	Anaemia in Thailand is linked to genetic factors like thalassemia, which can lead to physical symptoms like fatigue
Vietnam	Women of reproductive age and children under 5	23.1% (women), 23.3% (children)	Iron deficiency, micronutrient deficiencies, parasitic infections, genetic disorders, socioeconomic factors	Increased mortality, stunted development, cognitive decline	Nguyen et al. (2015)	Anaemia prevalence in Vietnam is influenced by multiple factors, which impact mortality and development in women and children

laboratory-based tests, but they are useful for screening and monitoring purposes in resource-limited areas (Sue Pavord et al., 2012).

Challenges in Haemoglobin Colour Scale

Among the above-mentioned methods, haemoglobin colour scale is a simple and low-cost method. However, there are many challenges with this method:

- Result interpretation: It can be difficult to accurately interpret results from the colour scale. This can happen especially in cases where the blood sample colour is quite close to the reference solution colour. In such cases, misdiagnosis of anaemia is possible.
- Inaccuracy in reading: The accuracy of this method is dependent on the preparation of the blood sample and colour matching. If the preparation is not proper and colour matching is not accurate, it will lead to inaccuracy in value reading (Perez-Plazola et al., 2020).
- Range limitation: The colour scale has only limited range. If the colour comes out of range, it can create false positives or false negatives.
- Light sensitivity: The colour scale can fade over time due to sensitivity to light. If the reference solution is not stored properly, it can give inaccurate values.

Training requirement: To work with a colour scale, healthcare workers need to be trained carefully. Without proper training and supervision, they will not be able to use the colour scale correctly. Although there are challenges in the haemoglobin colour scale method, it is still a low-cost, useful method where other methods are not available (Joseph et al., 2016).

Chances of Error in Haemoglobin Colour Scale

The use of haemoglobin colour scales for estimating haemoglobin levels can be prone to certain errors and limitations. Here are some of the potential sources of error:

- Lighting Conditions: The accuracy of the haemoglobin colour scale can be affected by variations in lighting conditions, such as differences in natural light or the use of artificial lighting.
- Colour Vision: Individuals with colour vision deficiencies may have difficulty distinguishing between certain colours on the haemoglobin colour scale, which can lead to inaccuracies in the estimated haemoglobin level.
- Interpretation Errors: The use of the haemoglobin colour scale requires interpretation by the user, which can lead to errors due to the differences in perception or understanding of the scale.
- Other Factors Affecting Skin Colour: Skin colour can be affected by a variety of factors, such as sun exposure, skin conditions, or the use of certain medications,

which can make it difficult to accurately assess haemoglobin levels using the colour scale.

- Variability in Skin Colour: Skin colour can vary widely across individuals and populations, which can make it difficult to develop a standardised colour scale that is accurate across all contexts.
- Despite these limitations, haemoglobin colour scales can be a useful tool for estimating haemoglobin levels in settings where more accurate methods of measurement may not be available or feasible. However, it is important to be aware of the potential sources of error and to interpret the results of the colour scale with caution (Van den Broek et al., 1999).

Available Non-invasive Method for Haemoglobin Measurement

There are non-invasive methods for measuring haemoglobin levels that are becoming popular, especially in regions where traditional blood tests may not be feasible or practical. Some available non-invasive methods for haemoglobin measurement include:

- Pulse oximetry: This is a non-invasive method that uses a sensor to measure oxygen saturation in the blood. While it does not directly measure haemoglobin levels, it can be used as a proxy, as low haemoglobin levels can cause a decrease in oxygen saturation.
- Near-Infrared Spectroscopy (NIRS): This is a non-invasive technique that uses light to measure the amount of haemoglobin in the blood. A light is shone through the skin, and the amount of light absorbed by the blood is measured. This method can measure both total haemoglobin levels and the levels of oxygenated and deoxygenated haemoglobin (Joseph et al., 2016).
- Photoplethysmography (PPG): This is a non-invasive method where changes in blood volume in the skin are measured. A light is shone onto the skin, and the amount of light reflected is measured. Haemoglobin level is estimated by assessing the amount of blood in the skin (Joseph et al., 2016).
- Haemoglobinometers: Some haemoglobinometers use a non-invasive method by shining a light on the skin and measuring the amount of light absorbed. This method is often used in settings where blood tests are not available.
- Although non-invasive methods for haemoglobin measurement are favourable, they are still in the development stage, and their accuracy may not match that of traditional blood tests. More research is needed to determine the accuracy and reliability of these methods across various populations and situations (Joseph et al., 2016).

WHO-Certified Haemoglobin Colour Chart Values Inaccuracy

The World Health Organisation (WHO) recommends the use of a haemoglobin colour chart as a rapid and accurate method of estimating haemoglobin levels in low-resource settings where laboratory testing is not available. The accuracy of the haemoglobin colour chart depends on several factors, including the quality of the chart, the lighting conditions under which it is used, and the skill of the user. The WHO has developed and validated several haemoglobin colour charts, including the HemoCue® Hb 201+ system and the Hemocue® B-Haemoglobin system. These charts have been extensively tested and are considered highly accurate for estimating haemoglobin levels in the range of 4–17 g/dL. However, it is important to note that the haemoglobin colour chart is not a substitute for laboratory testing and should be used only as a screening tool. If the haemoglobin colour chart indicates a low haemoglobin level, confirmatory laboratory testing should be performed to diagnose and manage the underlying condition (Darshana & Uluwaduge, 2014).

WHO Certified Haemoglobin Colour Chart Standard Deviation in Study

The World Health Organisation (WHO) has conducted several studies to evaluate the accuracy and reliability of various haemoglobin colour charts, including the HemoCue® Hb 201+ system and the HemoCue® B-Haemoglobin system. In these studies, the WHO has calculated the standard deviation (SD) of the haemoglobin colour chart readings as a measure of its precision. For example, in a study of the HemoCue® B-Haemoglobin system, the WHO reported an SD of 0.78 g/dL for readings taken by trained health workers and an SD of 1.06 g/dL for readings taken by untrained volunteers. The WHO has also established performance targets for haemoglobin colour charts, including a target of ± 1.0 g/dL for the absolute difference between the haemoglobin colour chart reading and the laboratory haemoglobin measurement. This performance target is used to evaluate the accuracy and precision of haemoglobin colour charts in different settings (Thomas et al., 2005). Overall, the WHO has found that haemoglobin colour charts are a reliable and accurate tool for screening and monitoring haemoglobin levels in low-resource settings, provided they are used correctly and in conjunction with confirmatory laboratory testing as needed.

Major Drawback in Existing Non-invasive Method for Haemoglobin Measurement

There are several limitations and drawbacks in existing non-invasive methods for haemoglobin measurement, including:

- Accuracy: Non-invasive methods are not as accurate as traditional blood tests. Their accuracy can be affected by factors such as skin pigmentation, age and other medical conditions (Srivastava et al., 2014).
- Cost: Some non-invasive methods can be expensive and require specialised equipment, such as near-infrared spectroscopy.
- Training: Healthcare workers may require specialised training to use these methods effectively and to interpret the readings correctly.
- Interference: External factors such as movement, ambient light and temperature can affect non-invasive methods for haemoglobin measurement and their accuracy can be compromised.
- Limited Range: Non-invasive methods have a limited range of measurement, which may not be suitable for all populations. For example, some methods may not be accurate for infants or elderly individuals (Rice et al., 2013).
- Availability: Non-invasive methods may be difficult to implement in all healthcare settings, especially in resource-limited areas. Despite these drawbacks, non-invasive methods for haemoglobin measurement are still useful for screening and monitoring haemoglobin levels in some populations. The accuracy and availability of non-invasive methods for haemoglobin measurement may be improved by future research (Dewhirst et al., 2014).

Measurement of Haemoglobin Using Colour of Tongue or Eye

It is not a reliable or accurate method to measure haemoglobin using the colour of the tongue. A pale or white tongue can be a sign of anaemia, but many other factors that can affect the colour of the tongue, such as poor oral hygiene, dehydration, or certain medications. Similarly, measuring haemoglobin levels using eye colour is also not a reliable or accurate method. While the conjunctiva (the thin layer of tissue covering the eye) can appear pale in people with low haemoglobin levels, many other factors can affect eye colour, such as genetics, age and medical conditions. Haemoglobin levels are best measured using laboratory tests or non-invasive methods such as pulse oximetry or near-infrared spectroscopy. These methods are far more accurate and reliable than assessing the colour of the eye or tongue (Das et al., 2022).

Using Artificial Intelligence and Machine Learning for Anaemia Detection

Artificial intelligence (AI) and machine learning (ML) algorithms play a crucial role in the identification and classification of various diseases and health issues (Pradeep Ghantasala et al., 2022; Sasubilli et al., 2020). These technologies are applied to identify and classify Anaemia by analysing haemoglobin levels. Recent advancements, including the use of special cameras or neural networks to analyse images, show promise in accurately detecting anaemia, offering affordable and user-friendly methods that could aid in early detection. Nevertheless, challenges persist, such as ensuring the applicability of these methods to diverse populations and integrating them into healthcare systems. Effective collaboration among scientists, doctors and policymakers is essential to harness the full potential of AI in combating Anaemia.

While machine learning approaches for measuring haemoglobin levels based on tongue colour have limited research, they present challenges due to factors like hydration and nutrition influencing the reliability of this method (Narayan et al., 2020). Researchers emphasise the need for a dataset of tongue images and corresponding haemoglobin levels to train the algorithm. Validation on new data is crucial to assess accuracy and reliability (Fuadah et al., 2020). However, using machine learning in medical diagnosis and treatment faces challenges like obtaining large, high-quality datasets, managing bias and error risks and interpreting results. Extensive research and validation are prerequisites to ensure the safety, effectiveness and reliability of such an approach. It's worth noting that, currently, laboratory tests and non-invasive methods like pulse oximetry and near-infrared spectroscopy remain the most reliable for measuring haemoglobin levels (Collings et al., 2016; Kumar et al., 2021).

In another study, Lee SY, Chen CM, Lim EY, Shen L, Sathe A, Singh A, et al. implemented a Convolutional Neural Network (CNN)-based algorithm for detecting HbH inclusions. Training involved digital images of HbH-positive and HbH-negative blood smears. The software demonstrated sensitivity and specificity close to 91% and 99%, respectively, for HbH+ cells at various magnifications. Testing the AI-aided diagnostic model on 40 whole slide images (WSIs) showed good inter-rater reliability and high sensitivity and specificity at the slide level. To reduce misclassification, analysing a minimum of 2.4×10^6 red cells is estimated, based on the lowest frequency of HbH+ cells observed. Validation on 78 image sets confirmed the minimum required smear size. The development of AI software applications involved deep learning techniques, specifically using Deep CNN with ResNet and RCNN with ResNet-50 feature extraction. Initialising the weights of the feature extraction backbone contributed to reducing training time (Lee et al., 2021) (Table 7.4).

Future Potential Study of Anaemia Detection

A haemoglobin map is a geographical map that displays the prevalence of anaemia or low haemoglobin levels in a particular region or population. Such a map could help identify areas with a high prevalence of anaemia, target interventions to these areas and monitor progress in reducing anaemia rates over time (Ghermandi, 2016). Geographic Information System (GIS) technology can be applied for public health monitoring and decision-making (Fletcher-Lartey & Caprarelli, 2016). To develop a haemoglobin map, data would need to be collected on the haemoglobin levels of individuals in the population of interest. This data could be collected through blood tests, non-invasive methods such as pulse oximetry or near-infrared spectroscopy, or other means. The data would then need to be analysed to determine the prevalence

Table 7.4 Overview of studies on machine learning models for anaemia detection: a comprehensive analysis of primary research

Title/motive	Method used	Evaluation metrics	Image augmentation	Data augmentation methods used	Origin of dataset	Size of dataset	Accuracy	Conclusion
Detecting anaemia from retinal images using deep learning (Agrawal, 2021)	CNN; architecture used: ResNet, MobileNetV2 and DenseNet	Precision, AUC, Weighted log loss metric	Yes	Rotation, zoom, translation (shift) and flipping	Sankara Nethralaya, Hospital, Chennai (India)	400	WLLM had a higher result of 74%	MobileNetV2 had a higher performance. AUC achieved 67.10% and a precision of 567.00%
A non-invasive computerised technique to detect anaemia using images of eye conjunctiva (Bauskar et al., 2019)	SVM	Accuracy, specificity and sensitivity	No	N/A	Maulana Azad National Institute of Technology, Bhopal (India)	99	93% accuracy was achieved	The SVM attained 93% accuracy
Neural network-based non-invasive method to detect anaemia from images of eye conjunctiva (Jain et al., 2020)	ANN	Accuracy, specificity, sensitivity and confusion metric	Yes	Mirroring, rotation and translation	Italy and India	99	97% accuracy was achieved	The ANN model detected anaemia with a significant accuracy of 97%
Convolutional neural network for anaemia detection based on conjunctiva palpebral images (Magdalena et al., 2022)	CNN; architecture used: AlexNet	Precision, recall, accuracy and F1-score	Yes	Not mentioned	Not indicated	2000	94% of accuracy	At the learning rate of 0.0001, the CNN had an accuracy of 94% by using Adam optimiser

(continued)

Table 7.4 (continued)

Title/motive	Method used	Evaluation metrics	Image augmentation	Data augmentation methods used	Origin of dataset	Size of dataset	Accuracy	Conclusion
Classification of anaemia with digital images of nails and palms using the Naive Bayes method (Peksi et al., 2021)	Naïve Bayes	Accuracy, specificity and sensitivity	No	N/A	Not indicated	20	An accuracy of 90% was achieved	The Naïve Bayes had an accuracy performance of 90%
Computer aided detection of anaemia-like pallor (Roychowdhury et al., 2017)	SVM, K-NN, DT, ANN	Accuracy, precision and recall	No	N/A	Not indicated	83	86% accuracy	86% classification accuracy was attained
Comparative study between Decision Tree, SVM and K-NN to predict anaemic condition (Noor et al., 2019a, 2019b)	MLR, DT, SVR	RMSE, LSVR, MLR	No	N/A	Chittagong Medical and Cox's Bazar College Hospitals (Bangladesh)	104	88.99% accuracy achieved	Decision Tree had a higher classification accuracy of 88.99%
Detection of anaemia from image of the anterior conjunctiva of the eye by image processing and thresholding (Tamir et al., 2017)	RGB thresholding algorithm	N/A	No	N/A	Dhaka Medical College (Bangladesh)	19	78.90% accuracy	The RGB thresholding algorithm used achieved an accuracy of 78.90%

(continued)

Table 7.4 (continued)

Title/motive	Method used	Evaluation metrics	Image augmentation	Data augmentation methods used	Origin of dataset	Size of dataset	Accuracy	Conclusion
An efficient technique of haemoglobin level screening using machine learning algorithms (Noor et al., 2019a, 2019b)	DT, SVM, K-NN	Confusion metric	No	N/A	Chittagong and Cox's Bazar Medical College Hospitals (Bangladesh)	104	Accuracy of 82.61%	A higher accuracy of 82.61 was attained by the Decision Tree algorithm
Anaemia detection using a full embedded mobile application with YOLO algorithm (Rivero-Palacio et al., 2021)	YOLO v5 Neural Network	Accuracy, recall, specificity and MAP confusion	Yes	Flipping and rotation	Universidad Peruana Cayetano Heredia (Peru)	457	93% accuracy	93% accuracy on PC and 80% on phone applications when integrated
Iron deficiency anaemia detection using machine learning models: a comparative study of fingernails, palm and conjunctiva of the eye images (Asare et al., 2023)	CNN, DT, SVM, Naïve Bayes, k-NN	Accuracy, f1-score, AUC, precision, recall	Yes	Flipping (mirroring), rotation and translation	Ten (10) selected hospitals across Ghana	527	An accuracy of 99.96% was achieved by the Naïve Bayes	The palpable palm images are significant for the detection of anaemia

(continued)

Table 7.4 (continued)

Title/motive	Method used	Evaluation metrics	Image augmentation	Data augmentation methods used	Origin of dataset	Size of dataset	Accuracy	Conclusion
An intelligent non-invasive system for automated diagnosis of anaemia exploiting a novel dataset (Dimauro et al., 2023)	SVM, random forest, k-NN, MobileNet-V2	Accuracy, sensitivity, specificity	Yes	Not indicated	Bari (Italy) and Karapakkam, Chennai (India)	123 (Italy) 95 (India) 218 (total)	91% accuracy achieved by the MobileNet-V2	The MobileNet-V2 achieved an accuracy of 91% on the Italy dataset, while the India dataset had an accuracy of 68%
A new artificial intelligence approach using extreme learning machine as the potentially effective model to predict and classify anaemia severity (Saputra et al., 2023)	ELM	Accuracy	Yes	N/A	Indonesia	34	Achieved an accuracy of 85.29%	ELM achieved a satisfactory accuracy of 85.29%

of anaemia or low haemoglobin levels in different areas or subpopulations. Other factors that could be included in the haemoglobin map might include information on the prevalence of risk factors for anaemia, such as poor nutrition, parasitic infections, or pregnancy, as well as information on the availability of interventions such as iron supplementation, deworming, or food fortification. Once the haemoglobin map is developed, it can be used to inform policy decisions, target interventions and monitor progress in reducing anaemia rates over time. The study could involve two components:

1. Collecting the geo-tagged images of tongues.
2. Applying machine learning algorithms to classify haemoglobin levels based on the tongue images.

After classifying the tongues images for haemoglobin level range, the result will be verified against data obtained from hospitals and laboratories.

Conclusion

In this review paper, the importance of haemoglobin level for the human health is explained, and it continues with how haemoglobin measurement is related to anaemia and other genetic blood diseases. Both invasive and non-invasive methods of measuring haemoglobin levels are also discussed. Moreover, the factors that can affect the accuracy of haemoglobin measurement are discussed in detail. As a futuristic direction for this health-related work, a haemoglobin map to indicate the haemoglobin level across different countries can be created. The innovation will lead to a technological solution to a fundamental problem affecting the population. It will reach the population via Government health services. The biggest contribution of this innovation is the creation of a haemoglobin map for the surveyed area. This innovation will bring a paradigm shift in the policy ecosystem. Government policy can be effectively implemented on the ground, and its impact can be tracked through the real-time map. This innovation will be a small step towards Digital and Healthy ASEAN.

References

Agrawal, A. (2021). *Detecting Anaemia from retinal images using deep learning.*

Ali, M., Amin, M. R., Jarl, J., Chisholm, N., & Saha, S. (2021). Maternal health status and household food security on determining childhood Anaemia in Bangladesh-a nationwide cross-sectional study. *BMC Public Health, 21*(1), 1581.

Asare, J. W., Appiahene, P., Donkoh, E. T., & Dimauro, G. (2023). Iron deficiency Anaemia detection using machine learning models: A comparative study of fingernails, palm and conjunctiva of the eye images. *Engineering Reports*, e12667.

Bauskar, S., Jain, P., & Gyanchandani, M. (2019). A noninvasive computerized technique to detect Anaemia using images of eye conjunctiva. *Pattern Recognition and Image Analysis, 29*, 438–446.

Collings, S., Thompson, O., Hirst, E., Goossens, L., George, A., & Weinkove, R. (2016). Non-invasive detection of Anaemia using digital photographs of the conjunctiva. *PLoS ONE, 11*(4), e0153286.

Darshana, L. G. T., & Uluwaduge, D. I. (2014). Validation of the WHO hemoglobin color scale method. *Anemia, 2014*(1), 531670. https://doi.org/10.1155/2014/531670

Das, S., Kesarwani, A., Kisku, D. R., & Dalui, M. (2022, Aug). Non-invasive haemoglobin prediction using nail colour features: An approach of dimensionality reduction. In *Intelligent computing theories and application: 18th international conference, ICIC 2022, Xi'an, China, August 7–11, 2022, proceedings, part I* (pp. 811–824). Springer International Publishing.

Dewhirst, E., Naguib, A., Winch, P., Rice, J., Galantowicz, M., McConnell, P., & Tobias, J. D. (2014). Accuracy of noninvasive and continuous haemoglobin measurement by pulse co-oximetry during preoperative phlebotomy. *Journal of Intensive Care Medicine, 29*(4), 238–242.

Dimauro, G., Griseta, M. E., Camporeale, M. G., Clemente, F., Guarini, A., & Maglietta, R. (2023). An intelligent non-invasive system for automated diagnosis of Anaemia exploiting a novel dataset. *Artificial Intelligence in Medicine, 136*, 102477.

Fletcher-Lartey, S. M., & Caprarelli, G. (2016). Application of GIS technology in public health: Successes and challenges. *Parasitology, 143*(4), 401–415.

Fuadah, Y. N., Sa'idah, S., Wijayanto, I., Patmasari, R., & Magdalena, R. (2020, Oct). Non invasive Anaemia detection in pregnant women based on digital image processing and K-nearest neighbor. In *2020 3rd International conference on biomedical engineering (IBIOMED)* (pp. 60–64). IEEE.

Ghermandi, A. (2016). Analysis of intensity and spatial patterns of public use in natural treatment systems using geotagged photos from social media. *Water Research, 105*, 297–304.

Goyena, E. A., Maniego, L. V., & Angeles-Agdeppa, I. (2020). Determinants of haemoglobin status in Filipino children and women population based on a nationwide survey. *Philippine Journal of Science, 149*, 731–746.

Horowitz, K. M., Ingardia, C. J., & Borgida, A. F. (2013). Anaemia in pregnancy. *Clinics in Laboratory Medicine, 33*(2), 281–291.

Jain, P., Bauskar, S., & Gyanchandani, M. (2020). Neural network based non-invasive method to detect Anaemia from images of eye conjunctiva. *International Journal of Imaging Systems and Technology, 30*(1), 112–125.

Jamnok, J., Sanchaisuriya, K., Sanchaisuriya, P., Fucharoen, G., Fucharoen, S., & Ahmed, F. (2020). Factors associated with Anaemia and iron deficiency among women of reproductive age in Northeast Thailand: A cross-sectional study. *BMC Public Health, 20*(1), 1–8.

Joseph, B., Haider, A., & Rhee, P. (2016). Non-invasive haemoglobin monitoring. *International Journal of Surgery, 33*(Part B), 254–257. https://doi.org/10.1016/j.ijsu.2015.11.048. ISSN 1743–9191.

Kaur, K. (2014). Anaemia 'a silent killer' among women in India: Present scenario. *European Journal of Zoological Research, 3*(1), 32–36.

Keokenchanh, S., Kounnavong, S., Tokinobu, A., Midorikawa, K., Ikeda, W., Morita, A., Kitajima, T., & Sokejima, S. (2021). Prevalence of Anaemia and its associate factors among women of reproductive age in Lao PDR: Evidence from a nationally representative survey. *Anaemia, 2021*(1), 8823030.

Kumar, R. D., Guruprasad, S., Kansara, K., Rao, K. R., Mohan, M., Reddy, M. R., Prabhu, U. H., Prakash, P., Chakraborty, S., Das, S., & Madhusoodanan, K. N. (2021). A novel noninvasive haemoglobin sensing device for Anaemia screening. *IEEE Sensors Journal, 21*(13), 15318–15329.

Lee, S. Y., Chen, C. M., Lim, E. Y., Shen, L., Sathe, A., Singh, A., Sauer, J., Taghipour, K., & Yip, C. Y. C. (2021). Image analysis using machine learning for automated detection of haemoglobin H inclusions in blood smears—A method for morphologic detection of rare cells. *Journal of Pathology Informatics, 12*, 18.

Magdalena, R., Saidah, S., Ubaidah, I. D. W. S., Fuadah, Y. N., Herman, N., & Ibrahim, N. (2022). Convolutional neural network for Anaemia detection based on conjunctiva palpebral images. *Jurnal Teknik Informatika (JUTIF), 3*(2), 349–354.

Margawati, A., Syauqy, A., Utami, A., & Adespin, D. A. (2023). Prevalence of Anaemia and associated risk factors among pregnant women in Semarang, Indonesia, during COVID-19 Pandemic. *Ethiopian Journal of Health Sciences, 33*(3).

Narayan, S. S., Arif, I., Shalu, H., & Kadiwala, J. (2020). *A smartphone based multi input workflow for non-invasive estimation of haemoglobin levels using machine learning techniques*. arXiv preprint arXiv:2011.14370.

Nguyen, P. H., Gonzalez-Casanova, I., Nguyen, H., Pham, H., Truong, T. V., Nguyen, S., Martorell, R., & Ramakrishnan, U. (2015). Multicausal etiology of Anaemia among women of reproductive age in Vietnam. *European Journal of Clinical Nutrition, 69*(1), 107–113.

Nik Rosmawati, N. H., MohdNazri, S., & Mohd Ismail, I. (2012). The rate and risk factors for Anaemia among pregnant mothers in Jerteh Terengganu, Malaysia. *Journal of Multidisciplinary Healthcare, 2*(150), 2161–2711.

Noor, N. B., Anwar, M. S., & Dey, M. (2019a, Nov). Comparative study between decision tree, SVM and KNN to predict anaemic condition. In *2019 IEEE international conference on biomedical engineering, computer and information technology for health (BECITHCON)* (pp. 24–28). IEEE.

Noor, N. B., Anwar, M. S., & Dey, M. (2019b, Dec). An efficient technique of haemoglobin level screening using machine learning algorithms. In *2019 4th International conference on electrical information and communication technology (EICT)* (pp. 1–6). IEEE.

Onyeneho, N. G., Ozumba, B. C., & Subramanian, S. V. (2019). Determinants of childhood Anaemia in India. *Scientific Reports, 9*(1), 16540.

Pavord, S., Myers, B., Robinson, S., Allard, S., Strong, J., Oppenheimer, C., & British Committee for Standards in Haematology. (2012). UK guidelines on the management of iron deficiency in pregnancy. *British Journal of Haematology, 156*(5), 588–600.

Peksi, N. J., Yuwono, B., & Florestiyanto, M. Y. (2021). Classification of Anaemia with digital images of nails and palms using the Naive Bayes method. *Telematika: Jurnal Informatika Dan Teknologi Informasi, 18*(1), 118–130.

Perez-Plazola, M. S., Tyburski, E. A., Smart, L. R., Howard, T. A., Pfeiffer, A., Ware, R. E., Lam, W. A., & McGann, P. T. (2020). AnemoCheck-LRS: An optimized, color-based point-of-care test to identify severe Anaemia in limited-resource settings. *BMC Medicine, 18*(1), 337. https://doi.org/10.1186/s12916-020-01793-6

Pradeep Ghantasala, G. S., Nageswara Rao, D., & Patan, R. (2022). Recognition of dubious tissue by using supervised machine learning strategy. In *Applications of computational methods in manufacturing and product design: Select proceedings of IPDIMS 2020* (pp. 395–404). Springer Nature Singapore.

Rahman, M. A., Rahman, M. S., Aziz Rahman, M., Szymlek-Gay, E. A., Uddin, R., & Islam, S. M. S. (2021). Prevalence of and factors associated with Anaemia in women of reproductive age in Bangladesh, Maldives and Nepal: Evidence from nationally-representative survey data. *PLoS ONE, 16*(1), e0245335.

Rice, M. J., Gravenstein, N., & Morey, T. E. (2013). Noninvasive haemoglobin monitoring: How accurate is enough? *Anesthesia & Analgesia, 117*(4), 902–907.

Rivero-Palacio, M., Alfonso-Morales, W., & Caicedo-Bravo, E. (2021, May). Anaemia detection using a full embedded mobile application with YOLO algorithm. In *IEEE Colombian conference on applications of computational intelligence* (pp. 3–17). Springer International Publishing.

Roychowdhury, S., Sun, D., Bihis, M., Ren, J., Hage, P., & Rahman, H. H. (2017, Feb). Computer aided detection of Anaemia-like pallor. In *2017 IEEE EMBS international conference on biomedical & health informatics (BHI)* (pp. 461–464). IEEE.

Saputra, D. C. E., Sunat, K., & Ratnaningsih, T. (2023, Feb). A new artificial intelligence approach using extreme learning machine as the potentially effective model to predict and analyze the diagnosis of Anaemia. In *Healthcare* (Vol. 11, No. 5, p. 697). MDPI.

Sari, P., Herawati, D. M. D., Dhamayanti, M., & Hilmanto, D. (2022). Anaemia among adolescent girls in West Java, Indonesia: Related factors and consequences on the quality of life. *Nutrients, 14*(18), 3777.

Sasubilli, S. M., Kumar, A., & Dutt, V. (2020). Machine learning implementation on medical domain to identify disease insights using TMS. In *2020 International conference on advances in computing and communication engineering (ICACCE), Las Vegas, NV, USA* (pp. 1–4). https://doi.org/10.1109/ICACCE49060.2020.9154960

Srivastava, T., Negandhi, H., Neogi, S. B., Sharma, J., & Saxena, R. (2014). Methods for haemoglobin estimation: A review of what works. *Journal of Hematology & Transfusion, 2*(3), 1028.

Sunuwar, D. R., Singh, D. R., Chaudhary, N. K., Pradhan, P. M. S., Rai, P., & Tiwari, K. (2020). Prevalence and factors associated with Anaemia among women of reproductive age in seven South and Southeast Asian countries: Evidence from nationally representative surveys. *PLoS ONE, 15*(8), e0236449. https://doi.org/10.1371/journal.pone.0236449

Tamir, A., Jahan, C. S., Saif, M. S., Zaman, S. U., Islam, M. M., Khan, A. I., Fattah, S. A., & Shahnaz, C. (2017, Dec). Detection of Anaemia from image of the anterior conjunctiva of the eye by image processing and thresholding. In *2017 IEEE region 10 humanitarian technology conference (R10-HTC)* (pp. 697–701). IEEE.

Thomas, L., Franck, S., Messinger, M., Linssen, J., Thomé, M., & Thomas, C. (2005). Reticulocyte haemoglobin measurement–comparison of two methods in the diagnosis of iron-restricted erythropoiesis. *Clinical Chemistry and Laboratory Medicine (CCLM), 43*(11), 1193–1202.

Van den Broek, N. R., Ntonya, C., Mhango, E., & White, S. A. (1999). Diagnosing anaemia in pregnancy in rural clinics: Assessing the potential of the haemoglobin colour scale. *Bulletin of the World Health Organization, 77*(1), 15.

Wieringa, F. T., Dahl, M., Chamnan, C., Poirot, E., Kuong, K., Sophonneary, P., Sinuon, M., Greuffeille, V., Hong, R., Berger, J., & Dijkhuizen, M. A. (2016). The high prevalence of anaemia in Cambodian children and women cannot be satisfactorily explained by nutritional deficiencies or haemoglobin disorders. *Nutrients, 8*(6), 348.

Win, H. H., & Ko, M. K. (2018). Geographical disparities and determinants of anaemia among women of reproductive age in Myanmar: Analysis of the 2015–2016 Myanmar Demographic and Health Survey. *WHO South-East Asia Journal of Public Health, 7*(2), 107–113.

Chapter 8
Reinforcement of IT Skills for Entrepreneurial Mindset of Business Students

Diroubinee Mauree-Narrainen

Abstract Youth is a very delicate segment of the demographics of a society. Appropriate approaches for capacity building and resource mobilization for the youth ensure a sustainable future generation. The Organization for Economic Co-operation and Development (OECD) Ministerial Council Meeting 2022 advocates three interesting policies for the next generation and sustainable transition, namely (i) youth participation, (ii) policy action for future generations and (iii) environment and the planet (C/MIN(2022)16). Realization of these suggestions is feasible through student-cantered and inquiry-based learning. The purpose of these adopted teaching approaches in this study was to reinforce the IT skills of business students through a capstone project identified by the student in a chosen Small and Medium Enterprise (SME). This teaching strategy was implemented for business students at the School of Business Management and Finance at the University of Technology, Mauritius, during the first lockdown of the COVID-19 pandemic in March, 2020. Students were evaluated on their learning performance through a 100% project-based assessment, which was based either on an existing Micro, Small and Medium Enterprise (MSME) family business or a future small business that the student aspired to set up after completing their studies. The objectives of this teaching approach were to (i) impart knowledge related to the technical IT concepts to the students, (ii) cultivate their creative and innovative skills needed for entrepreneurship, (iii) instil problem-solving skills and (iv) develop students' IT support skills for their family businesses. These objectives contribute to creating an enabling environment for youth empowerment in technological learning, innovation in local start-ups and prospects for job creation. The outcome was interestingly positive, resulting in higher student curiosity in the technological fields for businesses, their willingness to opt for IT jobs and their interest in pursuing advanced IT qualifications.

Keywords Entrepreneurial mindset · Information technology skills · SDGs

---

D. Mauree-Narrainen (✉)
School of Business Management and Finance, University of Technology, Mauritius, Port Louis, Mauritius
e-mail: d.mnarrainen@utm.ac.mu

X. Poshiwa et al. (eds.), *Technological Innovations for Sustainable Industrialization*,
https://doi.org/10.1007/978-981-96-7995-9_8

Abbreviations

ACW	Africa Code Week
ATM	Automated Teller Machine
CLEn	Community Learning Engagement
EFTPOS	Electronic Funds Transfer Point of Sales
IB	Internet Banking
IT	Information Technology
KM	Knowledge Management
LMS	Learning Management System
MB	Mobile Banking
MIS	Management Information Systems
OECD	Organization for Economic Co-operation and Development
NGO	Non-Governmental Organization
SBMF	School of Business Management and Finance
SDG	Sustainable Development Goal
UTM	University of Technology, Mauritius

Introduction

The Organization for Economic Co-operation and Development (OECD) Ministerial Council Meeting 2022 advocates three key policies for the next generation and sustainable transition, namely (i) youth participation, (ii) policy action for future generations and (iii) environment and the planet. The present challenge is to build a healthy, knowledgeable and skilful workforce, along with robust infrastructure to contribute to the development of the Knowledge Economy (The World Bank, 2006, 2007), which has the following pillars: (i) economic and institutional regimes, (ii) education and human resources, (iii) information infrastructure and (iv) innovation systems, as depicted in Fig. 8.1.

It is mandatory to groom students with skills such as critical thinking, creativity, communication and collaboration and prepare them to be global citizens, thereby creating the prerequisite for arousing students' needs and questions beyond teachers' provided curriculum (Saqlain, 2016). This leads to a more explorative approach in learning and thereby elucidates Goertz's (2015) concept of inquiry-based learning as a 'student-directed, interest-driven approach to acquiring new knowledge'. It is also worth noting that inquiry-based learning supports students' information literacy with emphasis on analysis, reflection and evaluation of information through asking questions and investigating real-life problems, enhances students' attitudes towards learning, develops self-directed learning and improves their knowledge (Hmutra, 2016). On the same line of thinking, Guido (2017) clarifies it as learning that allows students to build knowledge through exploration, experience and discussions. Bozkurt et al. (2013) describe this approach as rooted in the philosophical movement

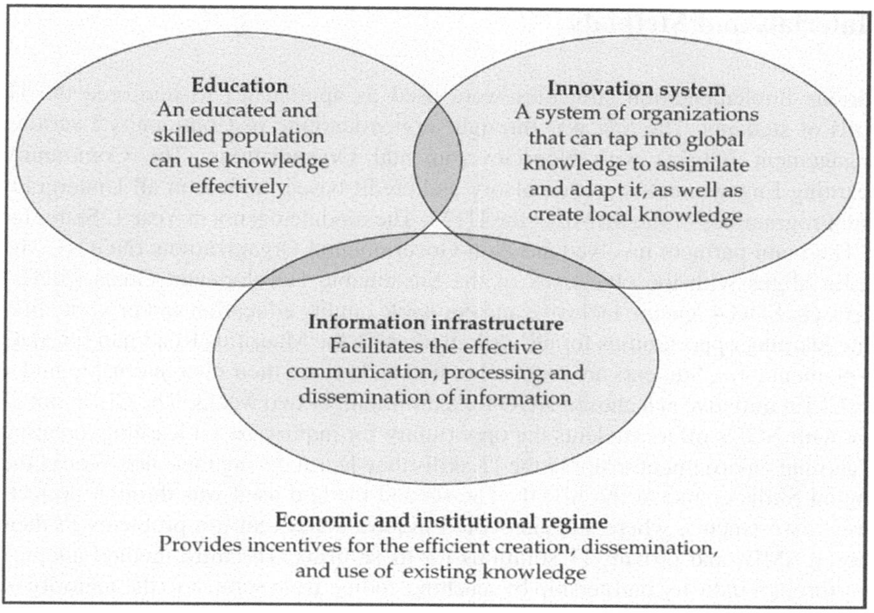

Fig. 8.1 Pillars of the knowledge economy. *Source* The World Bank (2007)

of pragmatism, while emphasizing the idea of being active and learning through research (Ormancı & Balım, 2019; Tatar, 2006). Thus, through teacher guidance, students are able to internalize acquired knowledge and skills (Tezel, 2018).

Once students have developed inquiry-based learning as their learning approach, they become self-dependent, which is an important trait of an entrepreneur. Entrepreneurship is both a life-long and life-wide experience and is thus the best means to combine learning with experience in official, educational endeavours (European Commission, 2011). To set the tone, an entrepreneurial mindset means that 'being an entrepreneur is a state of mind. It is about analysing the world and what is happening in terms of opportunities and possibilities, trying to understand how an individual intervention can enter the economic and social system in terms of construction and progress and putting into practice objectives and translating ideas into action' (Ferrero & Fioro, 2014). With reference to the policy framework for 2030 of the Ministry of Education (Alberta Education, 2010), education is to be transformed around key principles related to the three E's of twenty-first century education, namely (i) Engaged Thinker who thinks critically, uses technology to learn, innovate, communicate and discover, solves problems through multidisciplinary skills, is a life-long learner and adapts to changing circumstances; (ii) Ethical Citizen, who is humble, fair and open-minded, respectful, compassionate; has team spirit and connects with the community and world; (iii) Entrepreneurial Spirit, who is opportunistic, a hard worker, disciplined, strives for excellence, successful, competitive, adaptable, resilient, confident, a risk taker and a bold decision maker.

Materials and Methods

Various implementation strategies were used as approaches to reinforce the IT skills of students. The first was through service learning or Community Learning Engagement (CLEn) with Non-Governmental Organizations. The Community Learning Engagement is a compulsory and credit-based module in all Undergraduate Programmes of the SBMF at the UTM. The module occurs in Year 1, Semester 2. The main partners involved are Non-Governmental Organizations (NGOs). The CLEn aligns with the objectives of the Sustainable Development Goals (SDGs) precisely SDG 4 'ensure inclusive and equitable quality education and promote life-long learning opportunities for all' and aligns with the Mauritius Roadmap for SDG implementation. Students are required to find and secure their own internship under the CLEn initiative at a chosen NGO for a minimum of two weeks. The CLEn initiative with NGOs offers students the opportunity for inquiry-based learning, creating a learning environment to apply the IT skills they learnt during their first year in the Digital Skills course at the SBMF. The second method used was through project-based assessments, where students could inquire about real-life problems in their chosen SMEs and provide IT solutions for these firms. The third method adopted was through industry partnership by teaching coding from scratch to the majority of SBMF students studying IT modules in their programmes.

Results and Discussions

This section describes three implementation strategies used to support the reinforcement of IT skills for business students in developing an entrepreneurial mindset at the SBMF of the UTM.

Strategy 1: School of Business Management and Finance (SBMF) CLEn Unit Operational Activities

Having completed their CLEn initiative, students have developed their empathy, self-discipline, emotional intelligence as innate qualities, knowledge and skills, including IT skills, to support them in fulfilling their personal, professional, national and universal commitments and contribute to the student entrepreneurship mindset. It was observed that the majority of students, during their CLEn initiatives, made use of their IT skills to provide support as per the needs of the NGOs. Some of the IT skills included letter and report writing, graphic design, accounting tasks using spreadsheets, presentation design through PowerPoint, basic website development, amongst others. Hence, students confirmed that they had the opportunity to reinforce their skills learnt and put them into practice in this CLEn internship initiative.

Additionally, students have been encouraged to continue their community service engagement even after their CLEn initiative in Year 1, Semester 2.

Strategy 2: Kindle Student Entrepreneurship Mindset Through IT Project-Based Assessments

The second strategy was to foster student entrepreneurship mindset through IT project-based assessments. This initiative was introduced during the first lock-down of COVID-19 pandemic in March 2020. The aim was to provide guidance to SBMF students, who, in turn, support their family business and develop their ICT skills. During 2020 and 2021, assessments for IT-related modules were based on 100% project-based evaluation, which focused either on an existing MSME family business, or a future small business the student planned to establish after completing their studies, or an existing Non-Governmental Organization (NGO), or even a medium or large firm where working students were employed. IT modules at the SBMF are illustrated in Table 8.1.

The objectives of teaching and assessing the above subjects were to (i) impart knowledge related to technical IT concepts to the students, (ii) cultivate their creative and innovative skills needed for entrepreneurship, (iii) instil problem-solving skills and (iv) develop students' IT support skills for their family businesses. Additionally, the mentoring approach used included one-to-one supervision by the tutor to the

Table 8.1 IT subjects for SBMF programmes

No.	Programme	IT subject	Level
1	BSc (Hons) Accounting with Finance	Accounting information systems	Year 3 Semester 1
2	BSc (Hons) Human Resource Management	Human resource information systems	Year 3 Semester 1
3	BSc (Hons) Logistics and Transport Management	Transport and logistics information systems	Year 3 Semester 2
4	BSc (Hons) Banking and International Finance	Financial information systems	Year 2 Semester 2
5	BSc (Hons) Management programmes (streams: Law, General, Finance, Marketing)	Management information systems	Year 3 Semester 1
6	BSc (Hons) Management programmes (streams: Law, General, Finance, Marketing)	E-business	Year 2 Semester 1

student. Some students' testimonies regarding their qualitative experiences of the IT project-based assessments are detailed below.

Respondent 1 completed his B.Sc. (Hons) Management with Finance in October 2022 and commented that the modules E-Business and Management Information System (MIS) had helped him increase his understanding of using e-business solutions to attract more customers. His family business was able to shift from traditional payment methods to Electronic Funds Transfer at Point of Sale (EFTPOS) systems, e-payment and e-banking solutions. Subsequently, his family business experienced a reduction in costs and an increase in the quality and speed of service through Business Process Reengineering (BPR). The student also stated that his creativity and problem-solving skills improved. He was able to analyse the slow traditional Information Systems (IS) and replace them with new technological implementations. By using the software development life cycle model to create his business website, he was able to challenge his competitors by implementing a web application with minimum bugs and customer data protection. Before being introduced to the modules E-Business and MIS, he was unaware of how to manage a database or how to outsource in the context of a real business. After completing these two subjects, he was thus able to showcase his skills in terms of innovation, IT, problem-solving and creativity.

Respondent 2 accomplished his studies in B.Sc. (Hons) Management with a specialization in Marketing in April 2022 and shared that over the last 2 years, he was taught E-Business and MIS as important modules, which played an important role in how to manage a business effectively. The objective of the project assignment was to develop a new business and adopt appropriate IT solutions. He attempted to implement IT infrastructure in different forms to process data, collect information, store gathered materials, acquire knowledge and facilitate communication. 'As a result, my business gained a competitive edge in the market and could survive and thrive during the COVID-19 pandemic'. He gained insightful knowledge and information from the course on how to use different types of IT infrastructure to better manage his business and differentiate it from competitors.

Furthermore, respondent 3 completed her B.Sc. (Hons) Management with a specialization in Marketing in October 2022. She commented that E-Business and MIS modules helped her mother as a woman entrepreneur through the project-based assignments. She had the opportunity to develop her mother's business using the knowledge acquired from these modules and acquired many skills, such as IT proficiency. She redesigned her mother's business in a more professional way by developing its Facebook and Instagram pages on social media. She also developed an online portfolio for her mother to showcase her crafts online.

Likewise, respondent 4, who graduated with a B.Sc. Management specialization in Finance in November 2022, pointed out that the project-based assignment made her go beyond what was initially planned for her business. She started her business three years ago and has achieved great success in managing her business with the help of IT subjects and project-based assignments. These also helped her develop skills such as problem-solving, technical skills and the ability to create a digital promotional campaign for her products. Her belief was related to innovation and focus on quality products, sustained by a good information system, as the key to business growth and development.

Respondent 5 graduated in November 2022 with a B.Sc. Management stream, General. He is now a Business Analyst at Accenture Services. He stated: 'From 2019 to 2022, I completed the E-Business, MIS and Knowledge Management (KM) modules, which really helped me to develop these following skills namely creativity, analytical abilities, communication, organization, problem-solving, project management and time management. Through the knowledge acquired during this course, I have been able to develop a website and a database for the Non-Governmental Organization (NGO) for which I am a director. Throughout the various projects initiated in the E-Business, MIS and KM modules, I enhanced my leadership skills by guiding and coaching several junior students during their project assignments. Besides, the knowledge acquired during the course has been very relevant to my current job, and I am presently pursuing a Master of Computer Studies.'

Respondent 6, a final year student pursuing a B.Sc. (Hons) Management in Marketing, happily shared that he had the opportunity to gain an insight into the technical side of IT. He acquired basic knowledge on how systems function with their set of codebase, payments, cloud, servers and many software tools that allow enterprises to have a competitive edge over competitors. He was able to explore his presentation skills in PowerPoint with videos, music, animation, editing pictures and graphic design, which boosted his confidence level and improved his communication skills. He also developed his analytical skills by proposing a new business model integrated with IT; which pushed him to think out of the box with innovative and creative ideas to enhance the quality of his family business to a greater extent.

The above testimonies from six students who completed their study programmes demonstrate interesting outcomes from this strategic implementation, including strengthened bonding with parents and siblings, execution of business ideas generated amongst family members, creativity in businesses and the development of businesses through IT. This project-based assessment strategy is ongoing in the study programmes at the SBMF as part of the assessment for IT-related modules and paves the way towards contributing to SDGs 4, 8 and 9.

Strategy 3: Industry Partnerships

The third strategic implementation achieved was through industry partnership with the 3rd edition of the Africa Code Week (ACW), conducted at the beginning of November 2022 at the University of Technology, Mauritius, targeting business students at both undergraduate and postgraduate levels at the SBMF. This initiative was a digital literacy programme by the SAP Company, with the objective of instilling twenty-first century skills amongst youth across the continent of Africa. Acting as a driver of exchange, the ACW was organized by the SBMF in collaboration with the Rajiv Gandhi Science Centre (RGSC), the implementer of the ACW initiative in Mauritius. RGSC was supported by other partners, to provide nationwide training in 'Coding from Scratch' to students at primary, secondary and tertiary levels. This initiative contributed to achieving SDGs 4, 10 and 17. Around 355 students registered and completed the coding training through face-to-face and virtual modes. The participants responded very positively to their training and felt confident about adding this new coding skill learnt to their resumes.

Conclusions

It was possible to reinforce the IT skills for SBMF students and foster an entrepreneurial mindset through these three strategies implemented using inquiry-based learning to developed the twenty-first century skills needed for the youth in facing the challenges of the labour market. Positively, these students have develop focus in their career objectives and have been better prepared to contribute to Mauritius as a Knowledge Economy. The skills they developed during their CLEn experience also prepared them to be universal citizens, enabling them to adapt to international labour markets. The three strategic implementations at the SBMF contributed to achieving six SDGs: SDG 4 (Quality Education), SDG 8 (Decent Work and Economic Growth), SDG 9 (Industry, Innovation and Infrastructure), SDG 10 (Reduced Inequalities) and SDG 17 (Partnerships for the Goals). These objectives contributed to creating an enabling environment for youth empowerment in technological learning, innovation in local start-ups and prospects for job creation. The outcomes have interestingly led to increased students curiosity in technological fields for businesses, a greater willingness to pursue IT jobs and interest in advanced IT qualifications.

Acknowledgements (1) Organizers of the NAM S&T Centre. (2) Ministry of Education, Tertiary Education, Science and Technology, Republic of Mauritius [Office of Focal Point of NAM S&T Centre in Mauritius]. (3) School of Business Management and Finance (SBMF), University of Technology, Mauritius (UTM). (4) Rajiv Gandhi Science Centre (Mauritius) Partner for Africa Code Week.

References

Alberta Education. (2010). *Inspiring education: A dialogue with Albertans.* Edmonton.

Bozkurt, O., Ay, Y., & Fansa, M. (2013). The effects of inquiry-based learning on students'' science achievement and on their attitudes towards science and students'' opinions about the implementation of the method in the teaching process. *AbantİzzetBaysal University Journal of Faculty of Education, 13*(2), 241–256.

European Commission. (2011). *Entrepreneurship education: Enabling teachers as a critical success factor—A report on teacher education and training to prepare teachers for the challenge of entrepreneurship education.*

Ferrero, D., & Fioro, C. (2014). *Human spirits and entrepreneurship culture—A new perspective in the globalization era* [Kindle Edition]. Austin Macaulay.

Goertz, P. (2015). *What does a 21st century classroom look like: Inquiry*. Edutopia. Retrieved 7 July 2021, from https://www.edutopia.org/discussion/what-does-21st-century-classroom-lookinquiry.

Guido, M. (2017). *Inquiry-based learning definition, benefits & strategies* [Online]. Prodigy Math Blog. Available at: https://www.prodigygame.com/blog/inquiry-based-learning-definitionbenefits-strategies/

Hmutra, W. (2016). *Information literacy and inquiry-based learning* [Video]. Retrieved 7 July 2021, from https://www.youtube.com/watch?v=1DrzM9xqA_I

Ministerial Council Statement (June, 2022). "The Future We Want: Better Policies for the Next Generation And a Sustainable Transition". Meeting of the Council at Ministerial Level, pp. 9–10

Ormancı, Ü., & Balım, A. G. (2019). Inquiry-based learning approach. In A. G. Balım (Ed.), *Innovative approaches in science teaching* (pp. 9–44). Anı Publishing.

Saqlain, S. (2016). *Inquiry-based learning VS traditional learning* [Online] Available at: https://starryeyedkhanblog.com/2016/12/13/inquiry-based-learning-vs-traditional-learning-byahsan-saqlain

Tatar, N. (2006). *The effect of inquiry-based learning approach on scientific process skills, academic achievement and attitude in primary school science education*. Gazi University. Available at: https://tez2.yok.gov.tr

Tezel, Ö. (2018). Inquiry-based science teaching. In O. Karamustafaoğlu, Ö. Tezel, & U. Sarı (Eds.), *Activity supported science teaching with current approaches and methods* (pp. 74–96). Pegem Academy Publishing.

The World Bank. (2006). The Knowledge Economy, The KAM Methodology and World Bank Operations.

The World Bank, (2007). The International Bank for Reconstruction and Development.

Part III
Advanced Technologies, Infrastructure Development and Sustainable Industrialization

Chapter 9
S&T Capacity Building and Promotion of Innovation for Sustainable Industrialisation in India: Role of CSIR

Geetha Vani Rayasam and S. A. Hasan

Abstract Research and development (R&D) is essential for innovation and technological advancement, leading to socio-economic development and improvement of human well-being. It has the potential to address societal challenges such as healthcare, environmental sustainability, energy efficiency and many other critical needs. In the current global scenario, with the growing concerns about climate change and decreasing natural resources, it is imperative that growth is sustainable. This can only be tackled through a focus on research, which not only catalyses technological advancements but also nurtures a culture of innovation and scientific creativity, both of which are vital for long-term success and global competitiveness. The Council of Scientific and Industrial Research (CSIR), India, focuses not only on creating Science and Technology (S&T) led knowledge bases, products and technologies but also on human resource development in S&T and has immensely contributed to the socio-economic development of India since its establishment. This strategy has the potential to meet India's ambitious goals while balancing the pursuit of sustainable development goals with meeting the aspirations of its citizens on the other, through S&T led endeavours.

Keywords Development · Industrialization · Fellowship · Innovation · Research and development · Science and technology · Sustainable · SDG

Introduction

Sustainable development aims for progress and development that can meet existing requirements, without depriving future generations of the opportunity to fulfil their needs. While the concept of sustainable development can be interpreted differently,

G. V. Rayasam (✉) · S. A. Hasan
CSIR-National Institute of Science Communication and Policy Research (NIScPR), Dr KS Krishnan Marg, New Delhi 110 012, India
e-mail: gvrayasam@csir.res.in

X. Poshiwa et al. (eds.), *Technological Innovations for Sustainable Industrialization*,
https://doi.org/10.1007/978-981-96-7995-9_9

the aim remains that development should balance the needs of everyone without disregarding the environmental, social and economic constraints faced by people.

In the year 2015, leaders from 195 countries met at the United Nations General Assembly (UNGA) and agreed on a roadmap focused on ending poverty, protecting the planet and tackling inequalities. It was envisaged that progress could be achieved by establishing a close collaboration among national governments, industries and businesses, media organizations, academic and S&T institutions and local bodies, including Non-Governmental Organizations, to improve the lives of the people by 2030 (UN: https://sdgs.un.org/2030agenda). At the 2015 UNGA, the 2030 agenda for Sustainable Development was adopted, outlining the following 17 Sustainable Development Goals (SDGs) (UN: https://sdgs.un.org/2030agenda):

- SDG—1: No Poverty
- SDG—2: Zero Hunger
- SDG—3: Good Health and Well-Being
- SDG—4: Quality Education
- SDG—5: Gender Equality
- SDG—6: Clean Water and Sanitation
- SDG—7: Affordable and Clean Energy
- SDG—8: Decent Work and Economic Growth
- SDG—9: Industry, Innovation and Infrastructure
- SDG—10: Reduce Inequality
- SDG—11: Sustainable Cities and Communities
- SDG—12: Responsible Consumption and Production
- SDG—13: Climate Action
- SDG—14: Life Below Water
- SDG—15: Life on Land
- SDG—16: Peace, Justice and Strong Institutions
- SDG—17: Partnerships for Goals.

The Council of Scientific & Industrial Research (CSIR), India was established in 1942 to promote scientific and industrial research for the socio-economic progress of the country. CSIR is one of the world's major government supported research institutions and has played a prominent role in promoting technological innovation and entrepreneurship through its schemes and programmes on R&D and human resource development, which has contributed to industrial progress in India. At present, CSIR has 37 national research laboratories and 39 research centres spread across the country with 3476 active scientists supported by about 4000 technical and support personnel (CSIR: https://www.csir.res.in/about-us/about-csir). The research activities of CSIR cover almost all areas of science, engineering and technology including information technology and translational research. The role of CSIR is crucial for the country in the context of Vision 2047, which aims to make India a developed nation by 2047 (CSIR VISION, 2030).

The contributions made by CSIR in India's capacity building in science and technology, and to complementing the country's efforts towards industrial development, are discussed in this paper, along with the details of various schemes and programmes on R&D, and S&T human resource development.

Importance of R&D and Innovation for Sustainable Development

Research and Development (R&D) is a vital factor for an institution's ability for innovation. R&D activities lead to the development of new knowledge and technologies, processes, products and services to meet the needs of the consumers.

R&D plays a significant role in the development of sustainable solutions, which directly impacts the quality of life, and is an essential component of a nation's economic prosperity. Energy-efficient technologies, diversified farming, diagnostics and medicines, effective water supply network, wind and solar energy, waste utilization and management and environmentally friendly materials are just a few examples of R&D contributions to sustainability. The significance of R&D will continue to grow foresee ably. Technological progress and the prospects for finding new, novel and ideal solutions to global challenges will strengthen the role of R&D in building a sustainable future.

Now, as we have already passed halfway to 2030, scientists worldwide are increasingly in need of innovative solutions to advance the Sustainable Development Goals (SDGs) (DST: https://dst.gov.in/sites/default/files/STIP_Doc_1.4_Dec2020.pdf).

Progress towards some SDGs is stalled, especially in low-income countries, where the effects of the COVID-19 pandemic and ensuing crises have been severe (Gulnara & Ramona, 2024). Notwithstanding these challenges, the SDGs remain attainable and none of their targets are out of reach with active contributions from the S&T community through technological and innovative solutions (Gulnara & Ramona, 2024).

India's Success in S&T and Innovation, and Roadmap for Achieving SDGs

India is rapidly progressing in R&D and innovation, with nearly 100,000 patents granted in Financial Year (FY) 2023–24 (PIB: https://www.pib.gov.in/PressReleasePage.aspx?PRID=2034921), and achieved 31.6%growth in patent filings in 2022 (Tang et al., 2023), reflecting its advancing intellectual property portfolio. The country improved its position in the Global Innovation Index, rising to 40th rank in 2023 compared to 81st rank in 2015 (PIB, https://www.pib.gov.in/PressRelease). Total Ph.D. enrolment in India grew by 81.2% in 2021–22 compared to 2014–15 (from 117,000 in 2014–15 to 213,000 in 2021–22) (PIB: https://www.pib.gov.in/PressReleseDetailm.aspx?PRID=1999713). Further, the Gross Expenditure on Research and Development (GERD) has consistently grown, doubling from INR 60,196.8 crore (US$ 6950 million) in F.Y. 2010–11 to INR 127,381 crore (US$ 14,710 million) in F.Y. 2020–21.

India's position in high-quality research has significantly improved. According to the 2023 Nature Index (The Nature Index provides absolute Count and fractional Share counts of article publication at the institutional and national level and, as such, is an indicator of global high-quality research output and collaboration), India rose to 9th rank, surpassing Australia and Switzerland (Plackett, 2024) last four years—from 1039.7 in 2019 to 1394.7 in 2023 (Share is a metric that measures contribution to papers in the Index by authors based in a country compared to all authors of the article. For instance, if there are four authors of an article and all four are from India, it will add 1.0 to India's share. If two out of four authors are from India, it will add 0.5 to India's share).

India is a participant in the Global Pilot programme on Sustainable Development. As per its STI action plan, the country is actively pursuing four specific SDGs: (i) Zero Hunger (SDG-2); (ii) Good Health and Well Being (SDG-3); (iii) Clean Water and Sanitation (SDG-6) and (iv) Affordable and Clean Energy (SDG-7). These goals align closely with the 2030 Agenda, as they are interrelated with other UN-defined goals (UN: https://sdgs.un.org/sites/default/files/2023-10/STI%20for%20SDGs_RIS.PDF). Science, technology and innovation (STI) play a critical role in achieving these goals, which emphasise access, equality, inclusion and sustainability (UN: https://sdgs.un.org/sites/default/files/2023-10/STI%20for%20SDGs_RIS.PDF).

Recently, India has accelerated technology development and dissemination by establishing new partnerships with the private sector, including an expanding network of start-ups. While some fields have become self-sustaining, other areas are catching up rapidly. Government agencies and departments, including those focused on S&T, prioritize this approach. The Indian government launched the 'Accelerating Growth of New India's Innovations (AGNi)' programme, collaborating with key government departments to develop technological solutions for pressing challenges. This initiative aims to modernise R&D infrastructure for innovation by fostering linkages amongst innovators across industries, thereby facilitating the commercialisation of ground breaking solutions (UN: https://sdgs.un.org/sites/default/files/2023-10/STI%20for%20SDGs_RIS.PDF).

Schemes and Programmes of CSIR, India for Capacity Building in S&T and Innovation

The CSIR has envisaged the ***CSIR@2030 vision*** to improve the quality of life for citizens of India through innovative Science and Technology, and internationally competitive R&D, by evolving sustainable solutions and building capabilities so that India becomes self-reliant and achieves the dream of *Atmanirbhar Bharat* (self-reliant India) could be fulfilled (CSIR VISION, 2030).

Research is conducted at CSIR primarily under eight themes with multidisciplinary and trans-disciplinary approaches to meet its mandate. These themes are (CSIR: https://csirprojects.anusandhan.net/):

1. Aerospace, Electronics, Instrumentation and Strategic Sectors (AEIS);
2. Civil Infrastructure and Engineering (CIE);
3. Ecology, Environment, Earth, Ocean Sciences and Water (E3OW);
4. Mining, Minerals, Metals and Materials (4M);
5. Chemicals (including leather) and Petrochemicals (CIP);
6. Energy (conventional and non-conventional) and Energy Devices (EED);

7. Agriculture, Nutrition and Biotechnology (ANB);
8. Healthcare (HTC).

Research and Development activities at CSIR are categorised based on research focus, proposed outcomes and time frame. The projects aimed at translation and commercialization falls into the following categories:

Fast Track Translational (FTT) research projects aim to accelerate near-market, business-oriented projects across all technology spheres without thematic limitations. FTT projects emphasize national goals and address common societal problems. They focus on technology/product development with short-term outcomes (18 to 24 months). Over the last seven years, this initiative has successfully transferred numerous impactful technologies to the industry, demonstrating the value and effectiveness of the FTT program (CSIR: https://csirprojects.anusandhan.net/).

Fast Track Commercialization (FTC) research projects create novel commercial opportunities or expand existing commerce. These projects typically last 6–12 months and start with a low Technology Readiness Level (TRL). Multiple CSIR institutions/facilities collaborate on FTC projects (CSIR: https://csirprojects.anusandhan.net/).

Mission Mode Projects (MMPs) are research ventures with more ambitious and focused objectives and deliverables that are implemented in a specific area by synergising the best skills and capabilities existing in several CSIR labs and outside organizations. Further, these research projects have a significant emphasis on participation, aiming to develop pioneering, global standard and cost-justifiable technologies of high national significance. The projects' emphasis is on time-bound outcomes and deliverables, with 3 to 5-year duration. A few illustrations include programmes on Aroma, Phyto-pharmaceuticals, Cotton, Sickle Cell Anaemia, Chemical Intermediates, APIs and Catalysis (CSIR: https://csirprojects.anusandhan.net/).

New Millennium Indian Technology Leadership Initiative (NMITLI) is a flagship Program collaborating with industry partners for technology/product-oriented projects. Its goal is 'to catalyse innovation centred scientific and technological developments to help Indian industry achieve global leadership in selected niche areas, by synergizing the strengths of publicly funded R&D institutions, academia and private industry'. Both 'push' and 'pull' types of project proposals are advanced under this programme (CSIR: https://csirprojects.anusandhan.net/).

Focused Basic Research (FBR) projects are high-end basic science research projects, with early evidence (TRL 1–2). The typical duration is 12–24 months (CSIR: https://csirprojects.anusandhan.net/).

Niche Creating High Science/ High Technology (NCP) research ventures are built on niches established by CSIR groups to retain and achieve global leadership. The usual duration is 24 months (CSIR: https://csirprojects.anusandhan.net/).

Bilateral and Multilateral Projects (BMP) are aimed at reinforcing mutual R&D capabilities at the global level, implemented through cooperation modes (exploratory visits, joint meetings and workshops, joint research projects, setting up joint/associated laboratories with worldwide reputed institutions, including joint research and training centers). The international R&D collaboration projects are mostly focused on eight thematic approaches of CSIR (CSIR: https://csirprojects.anusandhan.net/) (Fig. 9.1).

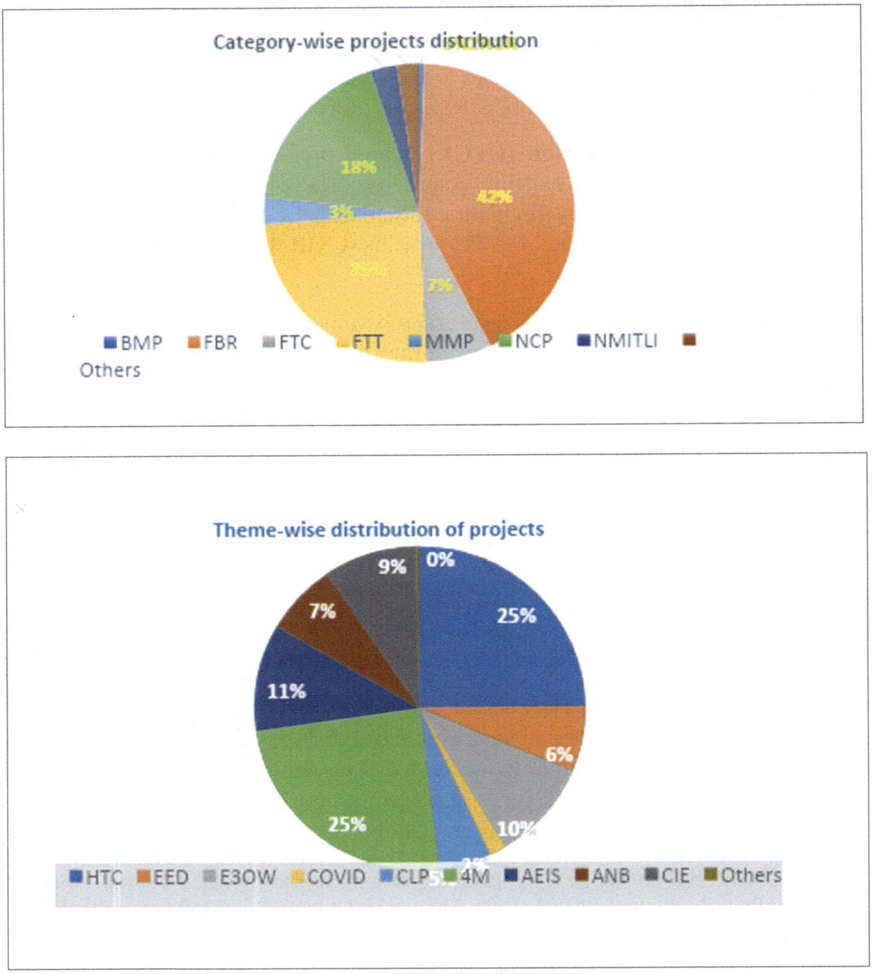

Fig. 9.1 Category-wise and theme-wise distribution of projects

Performance of CSIR

In 2022–23, CSIR published more than 5800 research papers (Fig. 9.2). Since its inception, CSIR has been granted more than 14,000 patents globally (CSIR: https://www.csir.res.in/about-us/about-csir; Wikipedia: https://en.wikipedia.org/wiki/Council_of_Scientific_and_Industrial_Research). During 2017–22, CSIR filed 1003 and 1056 patent applications in India and abroad, respectively (Fig. 9.3). A total of 1085 and 1443 patents were granted in India and abroad, respectively, during the same period (Fig. 9.4) (CSIR: https://www.csir.res.in/about-us/about-csir).

At present, CSIR has 900+ active patents across its eight theme areas. The distribution of active patents is as follows:

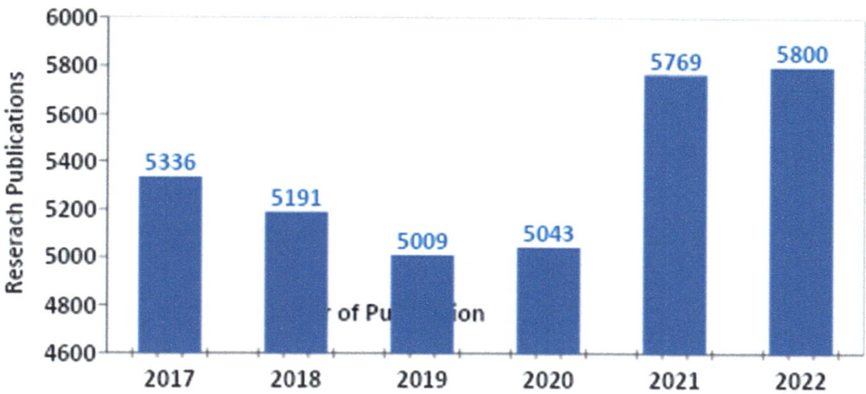

Fig. 9.2 Research publications from 2017 to 2022

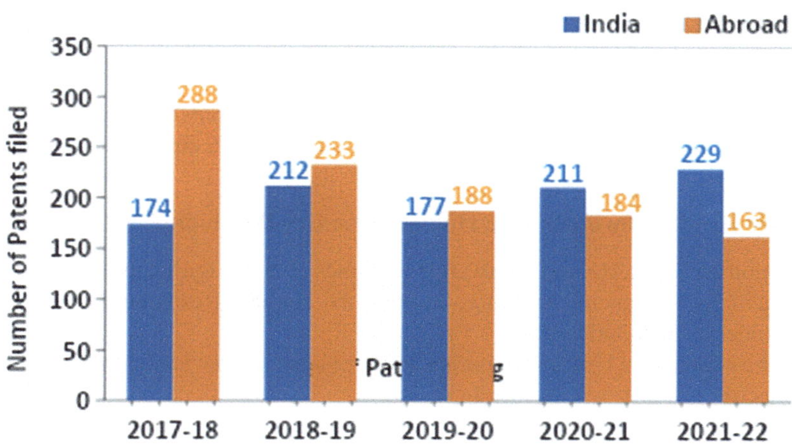

Fig. 9.3 Patents filed in India and abroad

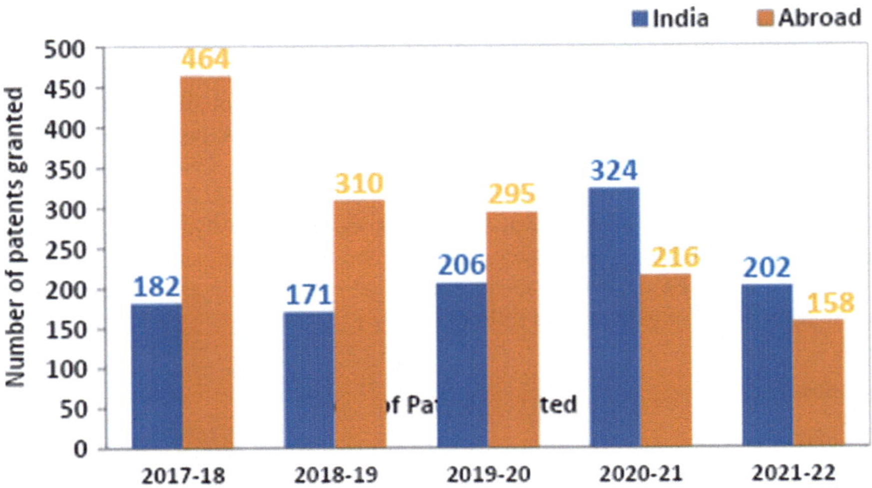

Fig. 9.4 Patents granted in India and abroad

At present, CSIR has **900+ active patents** across its eight theme areas. The distribution of active patents is as follows:

- Chemicals (including leather) and Petrochemicals: **333 (36.8%)**
- Healthcare: **183 (20.2%)**
- Agriculture, Nutrition and Biotechnology: **133 (14.7%)**
- Mining, Minerals, Metals and Materials: **73 (8.1%)**
- Aerospace, Electronics, Instrumentation and Strategic Sectors: **68 (7.5%)**
- Ecology, Environment, Earth/Ocean Sciences and Water: **53 (5.9%)**
- Energy (conventional and non-conventional) and Energy Devices: **34 (3.8%)**
- Civil Infrastructure and Engineering: **27 (3.0%)** (CSIR: https://techindiacsir.anusandhan.net).

Currently, CSIR has **1202 technologies** available for transfer at various Technology Readiness Levels (TRL 3 to TRL 9) (CSIR: https://techindiacsir.anusandhan.net).

Capacity Building through S&T Human Resource Development Programmes

In addition to its R&D and innovation schemes, CSIR significantly contributes to India's S&T capacity building through its Human Resource Development programmes, as detailed below.

CSIR established a Human Resource Development Group with the following aims and objectives (CSIR: www.csirhrdg.res.in):

1. To promote and foster the upgrading of a well-qualified pool of scientists, engineers and technologists for R&D in all disciplines of Science and Technology in the country.

2. To adopt a unified approach to S&T human resource development by inspiring and coordinating research in universities and research institutions.
3. To invest in basic science research and inter- and trans-disciplinary areas that are fore runners of 'high-tech' knowledge and expertise for the future.

Since its establishment, CSIR has implemented numerous fellowship programmes to enable enthusiastic youth to pursue research and innovation in science, engineering technology and trans-disciplinary areas. The primary objective of these fellowships is to identify brilliant young students with ground-breaking ideas, nurture their scientific aptitude and train them to conduct independent research. The key fellowship schemes implemented by Council in science, engineering and allied domains are discussed below.

1. **Doctoral and Post-doctoral Fellowships**

   At present, CSIR is supporting ~ 7600 doctoral and postdoctoral researchers pursuing research in about 700 R&D institutions, universities and colleges across the country.

   (a) **Doctoral Fellowships**

   Junior Research Fellowship (JRF-NET): The lead programme of its nature that offers scholarships to bright, budding youth in natural and interdisciplinary sciences. This National Eligibility Test (NET) is an all-India examination conducted twice a year for selecting Junior Research Fellows in five basic science areas: Chemical, Earth, Life, Mathematical and Physical Sciences. Selected candidates become eligible for a research fellowship for up to five years to pursue doctoral research leading to a PhD degree (CSIR: www.csirhrdg.res.in).

   Direct Senior Research Fellowship (SRF-D): It is for young researchers who do not need to qualify through national examinations and have enthusiasm for scientific and engineering research along with proven scientific credentials (e.g. research papers). A significant number of positions are awarded for research leading to a PhD (CSIR: www.csirhrdg.res.in). Each year, candidates with qualifications equivalent to JRF-NET and prior research experience are selected through interviews. During 2023–24, 362 SRF-Direct fellowships were awarded.

   (b) **Postdoctoral Fellowships**

   Research Associateship (RA): The fellowship is meant for post-doctoral research and each year a number of fellowships are awarded to those possessing the PhD degree in science or an *MD/MS/MDS/ME/MTech/Mpharm/MVSc* with three years of research experience. In 2023–24, 180 RA fellowships were awarded for postdoctoral research (CSIR: www.csirhrdg.res.in).

   Senior Research Associateship (SRA-erstwhile Pool officer): This is awarded to highly qualified academics and scientists, including Indians abroad to work in India (CSIR: https://www.csir.res.in/about-us/about-csir; CSIR: www.csirhrdg.res.in), who lack job opportunities in India. Researchers in India with outstanding academic records, who are unable to secure permanent jobs, are also eligible. Only bonafide Indian citizens may apply (CSIR: www.csirhrdg.res.in). Currently, CSIR supports approximately 78 SRAs.

Salient Features

Under the Doctoral and Postdoctoral Fellowship Programmes, a five-year age relaxation is granted to female students to encourage their pursuit of science and technology as a profession. As a result, the proportion of female recipients has increased in doctoral and postdoctoral fellowships. In 2017–18, women constituted 36% of all CSIR-funded doctoral and postdoctoral fellows, a figure that rose to 45% in 2023–24. Currently, out of 7645 research fellows, women research scholars constitute about 46.5% of the total researchers supported by CSIR. These research students are pursuing research in almost all the Indian states/UTs which have institutions of higher learning and research.

Additionally, since 2019, 24% (67) of the 275 Senior Research Associateships (SRA) were awarded to young Indian researchers previously working hard.

Since the programme's inception through 2023–24, approximately 170,000 doctoral and postdoctoral fellowships have been awarded (Fig. 9.5).

2. **Extramural Research Programmes**

   Apart from doctoral and postdoctoral fellowships to young researchers, CSIR offers research grants under the 'Research Scheme' to complement the current

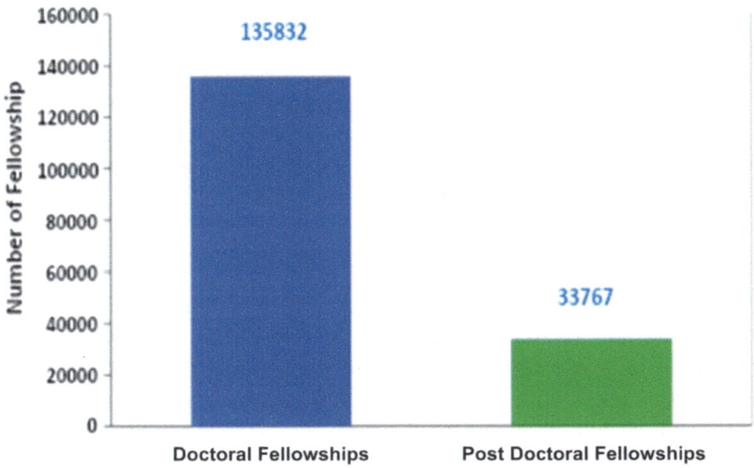

Fig. 9.5 Doctoral and post doctoral fellowships awarded since inception of the fellowship programmes

research work in academic and research institutions, etc. where basic research facilities exist (CSIR: www.csirhrdg.res.in). Under this scheme, financial grants are provided to support research work in the fields of Agriculture, Engineering and Technology, Medicine and Science. Research projects falling under basic sciences and applied nature are considered for research grants. Funding to the tune of INR 2.0 to 2.5 million is sanctioned in a project with at least one research fellow (JRF/SRF/RA), as well as contingency and equipment grants. State universities and scientists beginning their careers are among the beneficiaries of these projects (CSIR: https://www.csir.res.in/about-us/about-csir; CSIR: www.csirhrdg.res.in).

Funding research projects in universities/R&D institutions from all over India has encouraged young scientists and scientists from lesser-known places to pursue their scientific endeavours and has helped in building the R&D capabilities of these universities. Normally, every project has at least one research fellow (JRF/SRF/RA) working in it. In this way, the students are trained in various fields of S&T through these projects. This adds to the national S&T human resource development.

Since the inception of the Research Scheme in 1943, CSIR has sanctioned more than 11,500 research projects to universities, national R&D laboratories, colleges and institutions of national importance for carrying out high-end research and contributing towards the development of qualified human resources in science, engineering and technology.

3. **Emeritus Scientist Scheme**

The scheme is meant for retired, highly experienced and talented academics and researchers to perform state-of-the-art research and train fledgling scholars, investigators and academics. The emeritus scientists nurture and train young scientific talent to emerge as future scientists, publish their research findings in peer-reviewed journals that contribute to the world ranking of India for scientific publications, and in file patents to safeguard Intellectual Property Rights (IPRs).

The Emeritus Scientist programme plays a significant role in developing, nourishing and augmenting the stock of highly accomplished and competent science and engineering workforce in the country (CSIR: www.csirhrdg.res.in). Since the inception of the Scheme in 1958, CSIR has sanctioned more than 1100 Emeritus scientist projects to superannuated scientists.

4. **Promoting Knowledge Sharing**

To encourage interaction among scientists, inventors and domain experts, and the development of scientific temper, the Council offers monetary grants for holding national and international scientific seminars.

Further, CSIR provides travel grant for research scholars, and partial finance assistance to academics and researchers to present their research findings abroad with the purpose of enriching their knowledge through interaction with internationally acknowledged experts working in their research domains (CSIR: www.csirhrdg.res.in). It may be noted that during the last two years, i.e. 2022–23 and 2023–24, 525 and 723 proposals from young Indian researchers were sanctioned to present their research findings abroad before international audiences comprising subject specialists from all across the globe. This measure gives young researchers a great opportunity to get their research peer reviewed and learn about recent trends in their specific fields of research.

Since the inception of the Travel and Symposia Grant scheme, more than 28,000 proposals (15,166 for Travel Grants and 13,302 for Symposia Grants) have been approved, which has provided an excellent opportunity for young researchers to get exposure to current trends in science and technology and promote interaction between research scholars and international experts.

The number of Research Schemes sanctioned, Emeritus Scientists supported and Travel/Symposia Grants awarded since inception is given in Fig. 9.6. Further, in order to recognise excellence and outstanding research carried out by any citizen of India in science and technology, CSIR founded a *national-level prize* in the name of its founder, Shanti Swarup Bhatnagar (SSB). Since the inception of the SSB Prize in 1957 until 2022, a total of 597 scientists have been awarded this esteemed prize (CSIR: www.csirhrdg.res.in).

Also, in order to recognize brilliance and identify exceptional research work done by young scientists employed in the CSIR laboratories, the Council instituted the Young Scientist Award (YSA) in 1987 (CSIR: www.csirhrdg.res.in). Since its inception until 2022, a total of 221 scientists have been conferred with the YSA. The SSB Prizes and YSA awards have since been subsumed into the newly instituted 'Rashtriya Vigyan Puraskar' by the Government of India.

5. **Bhatnagar Fellowship**

   This fellowship leverages the expertise of outstanding researchers to develop novel technologies, enhance India's high-tech capabilities and guide academics

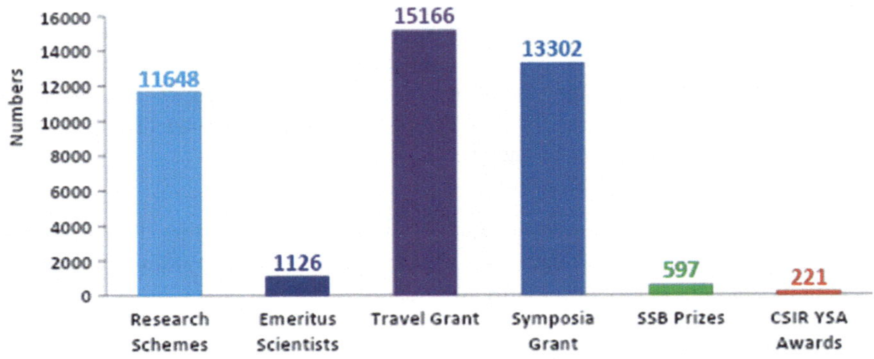

Fig. 9.6 Research scheme, emeritus scientists, travel grant, symposia grant, SSB prizes and YSA awards given since their inception

(CSIR: www.csirhrdg.res.in). From 1990 to 2021–22, 23, renowned scientists have been awarded the prestigious Bhatnagar Fellowship.

New Initiatives

(a) **Encouraging Female Researchers**

CSIR launched a programme exclusively for female researchers and academics to pursue high-end research in major disciplines of Science and Engineering (CSIR: https://es.csirhrdg.res.in/csir-aspire). Accordingly, 'A Special Call for Research Grants for Women Scientists (CSIR-ASPIRE)' was made (CSIR: https://www.csir.res.in/about-us/about-csir). The salient features of the CSIR-ASPIRE scheme are as follows:

- Eligibility: Only women scientists across the country are eligible to apply for research grants in five disciplines: Chemical Sciences, Engineering Sciences, Inter/Trans-Disciplinary Sciences, Life Sciences and Physical Sciences (CSIR: https://es.csirhrdg.res.in/csir-aspire).
- Funding: Total funds for each approved research project would be up to INR 3 million (CSIR: https://es.csirhrdg.res.in/csir-aspire).
- Duration: Projects have a maximum term of three years (CSIR: https://es.csirhrdg.res.in/csir-aspire).

A total of 2998 proposals were received, and discipline-specific search and selection committees sanctioned 301 proposals for support. The subject-wise data is given in the table below.

S. No.	Subject	Number of research proposals received	Number of research proposals sanctioned
1	Chemical sciences	349	34
2	Physical sciences	229	25
3	Engineering sciences	654	53
4	Inter/trans-disciplinary sciences	433	37
5	Life sciences	1333	152
	Total	2998	301

6. **Rashtriya Vigyan Puraskar (RVP): Honouring Scientific Excellence in India**

The Rashtriya Vigyan Puraskar (RVP) is one of the prestigious recognitions in the field of innovation, technology and science (PIB: https://www.pib.gov.in/PressNoteDetails.aspx?NoteId=152042&ModuleId=3. Established to acknowledge and inspire the remarkable contributions of scientists, technologists and innovators, the RVP honours those who have made path-breaking discoveries and advancements that significantly impact the society (PIB: https://www.pib.gov.in/PressNoteDetails.aspx?NoteId=152042&ModuleId=3). The RVP is awarded in following four distinct categories (PIB: https://www.pib.gov.in/PressNoteDetails.aspx?NoteId=152042&ModuleId=3):

1. Vigyan Ratna: It is the topmost honour, bestowed for lifetime accomplishments in any field of science and technology.
2. Vigyan Shri: Recognise notable contributions in any discipline of science and technology.
3. Vigyan Yuva-Shanti Swarup Bhatnagar: To honour young researchers who have made extraordinary contributions in science and technology.
4. Vigyan Team: Given to a team of three or more innovators/researchers/scientists who have contributed significantly in their area of research through collaborative work.

7. **Other Human Resource Development Programmes**

(a) **TWAS-CSIR Postgraduate Fellowship Program and Post-doctoral Fellowship Programme**

CSIR and UNESCO-TWAS have jointly instituted these fellowships for students from developing countries, who wish to pursue research in evolving areas of science and technology, leveraging the research infrastructure and facilities available at CSIR laboratories and institutes.

(b) **Raman Research Fellowship**

The Raman Research Fellowship is awarded to the scientists from CSIR for carrying out research in promising/high priority areas in reputed foreign R&D Centres of Excellence/institutions to augment their R&D skills/potentials (CSIR: https://www.csir.res.in/about-us/about-csir). The Fellowship is tenable abroad at CSIR cost and is given in all science disciplines. Since 2007–08 to 2023–24, 186 scientists from various CSIR labs have been awarded the CSIR-Raman Research Fellowship.

(c) **Integrated Skill Initiative Programme**

CSIR is contributing to the 'Skill India' programme of the Indian Government through its Integrated Skill Initiative. CSIR laboratories have organized numerous training programs for skill upgrading on pan-India basis to ensure employment generation and promotion of entrepreneurship. All such skill upgrading programs are consistent with the needs of the Indian industry. To realize the set goals, the programme aligns with the programmes initiated by the Ministry of Skill Development and Entrepreneurship. The programme helps in generating skilled manpower by providing and advancing their skills in all the doable fields (CSIO: www.csio.res.in).

8. **Human Resource Development Through the Academy of Scientific and Innovative Research (AcSIR)**

CSIR has established the AcSIR in 2011, now recognized as an 'Institution of National Importance'. Its mandate is to cultivate top scientific talent for future S&T research through innovative curricula, education and evaluation methods. AcSIR focuses on providing training and research opportunities in niche areas that are not routinely covered by Indian universities. It also offers joint academic programs with national and international educational institutions. Courses include Postgraduate Diploma, Master of Technology (M.Tech), Integrated Dual Degree Program (IDDP) and PhD in science and engineering. Currently, AcSIR has 2748 faculty members from CSIR laboratories and ~ 6500 students enrolled across its programmes (AcSIR: https://acsir.res.in/).

Role of CSIR in Promotion of Innovation and Industrialization

Over the past eight decades, CSIR has significantly contributed to India's industrialization, through R&D, technology development, product innovation, knowledge creation and human resource training. Many technologies suitable for industrial applications in critical areas have been developed by different laboratories of CSIR. Key technologies developed, transferred and commercialized by CSIR include.

Infant food: CSIR developed buffalo milk based infant food technology and transferred it to M/S Kaira District Milk Producers' Union (Amul), Anand, Gujarat. Commercialized as *Amul Spray* in the 1980s, this innovation made India become self-reliant in infant nutrition, reducing import costs and saving foreign exchange (CSIR: https://techindiacsir.anusandhan.net).

Health sector: CSIR developed *Centechroman*, a non-steroidal, once a week female oral contraceptive pill, licensed to M/S HLL Lifecare Ltd. The product was launched in the brand name *Saheli*. The Government of India incorporated Centechroman into the National Family Planning Program, distributing it free of cost to women under the brand name Chhaya (CSIR: https://techindiacsir.anusandhan.net).

CSIR developed Technologies for Natural Streptokinase, and Recombinant Streptokinase, transferring them to M/S Cadilla Pharma Ltd. and M/S Lupin Chemical Ltd. respectively. These firms successfully commercialised the products, enabling domestic production of these clot-dissolving drugs at significantly lower costs than imported alternatives (CSIR: https://techindiacsir.anusandhan.net).

CSIR developed affordable substitutes of anti-HIV drug cocktails to combat AIDS and transferred the technology to CIPLA. The company introduced these drugs in India and other developing countries at a fraction of the original cost, significantly improving access to life-saving treatment (CSIR: https://techindiacsir.anusandhan.net). Also, CSIR has played a vital role through its process technologies licensed to industries, which helped India become a leading generic drug supplier in the world.

Agriculture: The green revolution in India was pivotal for agricultural productivity, requiring both manpower and machinery. CSIR developed and introduced a 20-horsepower tractor named 'SWARAJ' and transferred the technology to a Public Sector Undertaking for production and marketing in 1974 (CSIR: https://www.csir.res.in/csir-milestones#). The tractor facilitated agricultural mechanization, and today, thousands of such tractors are used by Indian farmers for ploughing fields (CSIR: https://www.csir.res.in/csir-milestones#). Later, CSIR also developed a 60-horsepower tractor—'SONALIKA' (CSIR: https://www.csir.res.in/csir-milestones#).

The Indian government sought modest pumps suitable for villages without electricity, emphasising ease of operation and maintenance. CSIR developed inexpensive pumps, and approximately 3 million units are now deployed in rural India and third world countries (CSIR: https://www.csir.res.in/csir-milestones#).

The Indian Green Revolution initially relied on imported pesticides. Realizing this dependency, CSIR initiated a programme to develop economically viable processes for producing essential pesticides. Several pesticide technologies were developed and transferred to industries, and at one point, over 70% of India's pesticide production was based on CSIR-derived processes (CSIR: https://www.csir.res.in/csir-milestones#).

In the area of agro-technology, high oil yielding, pest-resistant varieties of menthol mint (*Menthaarvensis*) are developed, and India has now become the top producer and exporter of mint oil and its by products. These high oil yielding varieties have significantly contributed to the socio-economic development of farmers, particularly small and marginal farmers in rural India (CSIR: https://www.csir.res.in/csir-milestones#).

Kangra Valley in India is known for the cultivation and production of green tea, but plantations and production had started declining. To resuscitate and invigorate the plantations, CSIR invented and adopted innovative agro technologies and harvesting practices. These measures significantly helped increase the production of premium tea (CSIR: https://www.csir.res.in/csir-milestones#).

CSIR used DNA fingerprinting to create a database for Basmati rice to protect its intellectual property rights. Additionally, DNA fingerprinting was widely used in wildlife management and plant variety identification (CSIR: https://www.csir.res.in/csir-milestones#).

Chemicals, Minerals and Leather Sectors: Catalysts are the core of multi-billion dollar businesses, and multinational companies have been the frontrunners of this important segment. CSIR developed an inexpensive, long lasting zeolite technology and successfully transferred it to multinationals (CSIR: https://www.csir.res.in/csir-milestones#).

In the 1970s, India's chloralkali industry was heavily dependent on metal and graphite anodes and suffered from the recurrent replacement of anodes owing to their dimensional instability. CSIR developed an innovative solution: the titanium substrate insoluble anode (TSIA). The anode provides power savings of more than 12 per cent. Within a decade, the entire chloralkali industry in India shifted to TSIA. This CSIR innovation saved more than five billion Kilowatt-hours power (CSIR: https://www.csir.res.in/csir-milestones#).

The leather industry occupies a distinct position in the Indian economy, as it has massive potential for employment generation and economic growth. The leather and footwear industry engages about 4.5 million people, with a sizeable (about 30%) representation of women employees (CSIR: https://www.csir.res.in/en/csir-success-stories/leather-employment).

CSIR has developed numerous technologies, such as, the dehydrated chrome tanning technique, electro-oxidation-based zero effluent discharge, smart leathers, manure for agricultural use, dry tanning products, high-value collagenous goods, activated carbon from trimming and fleshing wastes, preservation-cum-unhairing process and biogas for energy conservation (CSIR: https://www.csir.res.in/en/csir-success-stories/leather-employment). The transfer of such technologies to the leather and leather product industries has increased both existing employee strength and the creation of new jobs.

In the mining industry, roof support systems designed to enhance worker have been developed by CSIR. Several mine scale fabricators are involved in the production of these roof systems (CSIR: https://www.csir.res.in/csir-milestones#).

India has vast coal reserves, which differ significantly in quality. CSIR know-how has facilitated the improvement of low grade coal for effective usage, with dozens of coal washeries in India upgrading tens of millions of tons of coal into higher-value, low sulphur coal for use in thermal power plants and steel industry, thereby reducing dependency on expensive and imported coal (CSIR: https://www.csir.res.in/csir-milestones#).

The diverse technologies of CSIR, spanning from solvent extraction to catalysts, have greatly impacted oil refineries in India and contributed to producing millions of tons of refined petroleum products. One of the oldest Indian refineries in the Assam was revitalized with the help of advanced molecular refinement technology developed by CSIR (CSIR: https://www.csir.res.in/csir-milestones#).

Aerospace Industry: CSIR successfully designed and manufactured India's first all-composite two seater aircraft, HANSA, and tested its initial flight in 1993 (CSIR: https://www.csir.res.in/csir-milestones#). CSIR also manufactured SARAS, a 14–18 seater multirole aircraft (CSIR: https://www.csir.res.in/csir-milestones#).

Supercomputing: CSIR connected several serial computers in parallel and transformed them into a supercomputer named as Flosolver, the first parallel computer built in India in 1986. This innovative transformation spurred other parallel computing missions in the country (CSIR: https://www.csir.res.in/csir-milestones#).

Contributions in Sustainable Industrialisation

Though industrialisation leads to economic prosperity and contributes to social development, it has adverse impacts on the environment, which pose a challenge to the entire ecosystem of the country. It is therefore necessary to strike a balance between meeting human economic and social needs and ensuring environmental protection. Sustainable industrial development is such a process that minimizes its environmental footprint while preserving economic advancement, social progression and quality of life (Jie et al., 2023). In recent times, due to the growing concerns about climate change and an increasing emphasis on sustainable development, efforts are underway by CSIR towards sustainable industrialisation while developing new technologies for commercial applications. A few recent significant initiatives undertaken by CSIR in this direction are:

Sustainable Fuel: Bio-jet fuel produced from Jatropha oil has been demonstrated in trial flights, which is a significant landmark, as it could be a sustainable substitute for conventional aircraft fuels. This bio-jet fuel technology from CSIR has formally been accepted for use in Indian Air Force Aircraft.

Additionally, Hydrogen Fuel Cell Technology was developed by CSIR-National Chemical Laboratory (CSIR-NCL), Pune, and CSIR-Central Electrochemical Research Institute (CSIR-CECRI), Karaikudi, Tamil Nadu, in collaboration with Sentient Labs, an R&D innovation lab incubated by KPIT Technologies and has been demonstrated to run cars and buses.

Storage Batteries and Clean Energy

Lithium-ion batteries have applications in energy storage systems, and CSIR has been working on the production of indigenous Li-ion batteries to reduce India's import dependence. An MoU was signed between CSIR-CECRI and Raasi Solar Power Pvt Ltd. The indigenous production of Li-ion batteries will greatly boost two flagship programmes of India: increasing the share of clean energy in the energy basket by generating 175 GW (of which 100 GW will be from solar energy) and the National Electric Mobility Mission to transition completely to electric vehicles by 2030 (CECRI: https://www.cecri.res.in/Portals/0/MoU_Rassi_CECRI.pdf).

Sustainable Agriculture Projects

Innovative programmes, such as the Aroma Mission, Floriculture Mission, Purple Revolution in Jammu & Kashmir, genome edited crops, transgenic cotton and trait-specific improvement in futuristic crops have been initiated by CSIR to raise farmers' income and promote entrepreneurship. These initiatives aim to transform farmers' lives by boosting local economies, creating employment and reducing urban migration. Many women have benefited from these schemes and became entrepreneurs.

Additionally, the Indigenous Compact Electric tractor (CSIR PRIMA ET11) for small marginal farmers offers an alternative to conventional diesel-powered systems, contributing to India's economic growth by providing an alternative to conventional diesel engine operated systems. This development supports MSMEs in entering the e-tractor segment, reduces pollution and noise in the villages and improves quality of life.

Environment Friendly Technologies and Products

CSIR developed 'Green Crackers' to control air pollution, a major health risk and transferred the technology to firecracker manufacturers (CSIR: https://techindiacsir.anusandhan.net).

CSIR has also made significant contributions to environmental remediation: An eco-rejuvenation technology (ERT) has been developed that revived a wasteland filled with fly ash into dense bamboo vegetation. The ERT is a potential tool for sustainable rural development through native biodiversity development on degraded land.

Further, manganese waste piles were turned into a lush green forest, and a coal mine spoil dump was transformed into a water body for aquaculture. Eroded lands have also been transformed into valuable assets at many places across the country (CSIR: https://www.csir.res.in/csir-milestones#).

Waste to Wealth

CSIR under the Govt. of India *Waste to Wealth* Programme, is exploring the use of steel slag as an alternative to natural aggregates in road construction as steel slag is one of the major solid wastes generated in the steel industry (CRRI: https://crridom.gov.in//node/1195). Approximately 0.1 million tons of processed steel slag aggregates were used in constructing a 1.2 km long six-lane bituminous road. Utilising steel slag ill reduces unsustainable quarrying of natural aggregates and cuts greenhouse gas emissions linked to road construction activities (CRRI: https://crridom.gov.in//node/1195). This green initiative supports India's commitment to the United Nation Sustainable Development Goal for 'Building Resilient Infrastructure through inclusive and sustainable industrialisation and green technologies' (CRRI: https://crridom.gov.in//node/1195).

Many other waste-to-wealth processes and technologies have been developed. Recently, CSIR has been working on *Vegan Leather*, wherein a leather-like product is made from mango pulp to make bags, belts, etc. This product is environmentally friendly substitute for conventional leather. CSIR has also developed technologies for generating wealth from waste, which include combining banana fibres with leather scrap, and producing exotic leather products from chicken feet skins (poultry waste) (CSIR: https://www.csir.res.in/en/csir-success-stories/leather-employment).

The high-rate biomethanation technology-based Anaerobic Gas lift Reactor (AGR) setup at the vegetable market yard is another notable example.

A technology for converting pathogenic biomedical waste into value-added soil additives has also been developed. A few technologies have been developed for converting plastic waste into floor tiles. These tiles have high density, are almost as tough as concrete, and yet flexible enough for Indian roads. 'Lead Free X-ray shielding red mud tiles', which are used for X-ray shielding on walls, floors and ceiling for CT scanner rooms, movable X-ray shields for operators, makeshift X-ray diagnostic centres, etc. are a few other innovations in the area of waste-to-wealth. Many CSIR labs are also working on processes for converting solid waste to energy and for electronic waste recycling, among others.

There are many such technologies in the offing, and CSIR laboratories are carrying out R&D towards the goal of sustainable industrialisation.

Conclusion

India has one of the youngest populations in an aging world. India's all-round progress depends on advancements in scientific R&D, innovation, technological development and expanding the S&T knowledge base. As illustrated through select examples of technologies developed, CSIR has contributed extensively to the nation's industrial development. With its recent emphasis on technological innovations for sustainable industrialisation across sectors, CSIR is positioned to complement the Government of India's efforts towards achieving the Sustainable Development Goals (SDGs).

CSIR is an active partner and contributor to many Government missions, such as Hydrogen mission, Carbon Capture, Utilization and Storage (CCUS) Mission, Quantum Mission, Sickle Cell Anaemia Mission and Biotechnology Mission, among others. These initiatives have positive impacts on sustainable industrialisation and achieving the Sustainable Development Goals (SDGs).

How CSIR responds to the future challenges and contributes to India's goal of becoming a developed country by 2047 will depend on the quality and quantity of trained scientific human resources. The various schemes CSIR operates for human resource development—through fellowships and grants have undoubtedly strengthened the country's achievements in the area of Science and Technology (S&T). CSIR plays a pivotal role in developing S&T human resources, significantly increasing the number of researchers per million population in India. According to World Bank data, India now has 260 researchers per million people, compared to 218 in 2015 and 110 in 2000. Recent improvements in global innovation rankings and India's position among the top three in research publications globally are promising indicators. With its demographic dividend, emphasis on S&T and focus on technopreneurship, India has the potential to become a leading innovation powerhouse in the coming years.

Acknowledgements The authors sincerely acknowledge all the present and past Directors-General of CSIR Heads of HRDG, and current and former members of CSIR-HRDG and various CSIR labs whose work is featured here. We also extend gratitude to other relevant divisions of CSIR, such as ISTAD, CSIR-HRDC and AcSIR for their valuable efforts and contributions over the years, which have been summarized in this article.

References

About CSIR, CSIR website, downloaded on 14 February 2025. https://www.csir.res.in/about-us/about-csir

AcSIR website, downloaded on 14 February 2025. https://acsir.res.in/

CECRI website, downloaded on 14 February 2025. https://www.cecri.res.in/Portals/0/MoU_Rassi_CECRI.pdf

CRRI website, downloaded on 14 February 2025. https://crridom.gov.in//node/1195

CSIO website downloaded on 14 February 2025. www.csio.res.in

CSIR ASPIRE Web portal referred on 14 February 2025. https://es.csirhrdg.res.in/csir-aspire

CSIR India R&D Projects, CSIR website downloaded on 14 February 2025. https://csirprojects.anusandhan.net/

CSIR Innovations, CSIR website, downloaded on 14 February 2025. https://www.csir.res.in/csir-milestones

CSIR Technology Showcase, CSIR website, downloaded on 14 February 2025. https://techindiacsir.anusandhan.net

CSIR VISION 2030 Document, CSIR website, downloaded on 14 February 2025.

CSIR-HRDG website downloaded on 14 February 2025. www.csirhrdg.res.in

Gulnara, D., & Ramona, T. (2024). Role of innovation in achieving the sustainable development goals: A bibliometric analysis. *Journal of Innovation & Knowledge, 9*, 1–13.

https://en.wikipedia.org/wiki/Council_of_Scientific_and_Industrial_Research
Jie, H., Khan, I., Alharthi, M., Zafar, M. W., & Saeed, A. (2023). Sustainable energy policy, socio-economic development, and ecological footprint: The economic significance of natural resources, population growth, and industrial development. *Utilities Policy, 81*. https://www.sciencedirect.com/science/article/pii/S0957178723000024?via%3Dihub
Leather for Employment, CSIR website, downloaded on 14 February 2025. https://www.csir.res.in/en/csir-success-stories/leather-employment
Plackett, B. (2024, June 18). *Nature Index News*. https://www.nature.com/nature-index/news/nature-index-research-leaders-india-follows-china-footsteps
Press Information Bureau (PIB) release of AISHE report, dated 25 January 2024. https://www.pib.gov.in/PressReleseDetailm.aspx?PRID=1999713
Press Information Bureau (PIB) release posted on 22 July 2024. https://www.pib.gov.in/PressReleasePage.aspx?PRID=2034921
Press Information Bureau (PIB) release posted on 28 September 2023. NITI Aayog, Government of India. https://www.pib.gov.in/PressReleasePage.aspx?PRID=1961576
Press Information Bureau (PIB) release, posted on 23 August 2024. https://www.pib.gov.in/PressNoteDetails.aspx?NoteId=152042&ModuleId=3
Research and Innovation System for Developing Countries (RIS) report on United Nations website downloaded on 14 February 2025. https://sdgs.un.org/sites/default/files/2023-10/STI%20for%20SDGs_RIS.PDF
Science, Technology and Innovation Policy, 2020, DST website-downloaded on 14 February 2025. https://dst.gov.in/sites/default/files/STIP_Doc_1.4_Dec2020.pdf
Tang, D., et al. (2023). *World intellectual property indicators 2023* (p. 30). WIPO, exhibit A.18.
United Nations website-downloaded on 14 February 2025. https://sdgs.un.org/2030agenda

Chapter 10
Sustainable Industrialisation in Agro-processing: Strategies, Innovations and Impacts

Xavier Poshiwa

Abstract Sustainable industrialisation in agro-processing is essential for transforming raw agricultural products into high-value goods while minimising environmental impact and enhancing economic viability. For most African economies, this holds promise, as in many currently industrialised economies, agro-processing has served as an entry point for industrialisation. The major strengths of many African countries include the availability of arable land, access to water, favourable climatic conditions and a youthful labour force that enables the cultivation and production of numerous agricultural commodities. This implies that Africa has the potential to support a robust agro-processing industry, as its outputs can be sold to other sectors as intermediate inputs for value addition. However, currently, value addition remains weak, and agricultural products are sold in domestic or international markets mostly as unprocessed commodities. Therefore, agro-processing in Africa remains underexploited and offers numerous opportunities for industrialisation. This paper explores the principles of sustainable industrialisation within the agro-processing sector, emphasising its challenges, strategies, technological innovations and their impacts on environmental and economic sustainability.

Keywords Agro-processing · Economic growth · Sustainable industrialisation · Technological innovations

Abbreviations

AI	Artificial intelligence
FAO	Food and Agriculture Organization
GATT	General Agreement on Tariffs and Trade
ICT	Information & Communication Technologies

X. Poshiwa (✉)
Great Zimbabwe University, Masvingo, Zimbabwe
e-mail: xposhiwa@gzu.ac.zw

X. Poshiwa et al. (eds.), *Technological Innovations for Sustainable Industrialization*,
https://doi.org/10.1007/978-981-96-7995-9_10

IoT	Internet of Things
4IR	Fourth Industrial Revolution
R&D	Research and Development
SDG	Sustainable Development Goal
SME	Small-scale and medium enterprises
SSA	Sub-Saharan Africa
STI	Science, technology and innovation
STEM	Science, technology, engineering and mathematics
TVET	Technical and vocational education and training

Introduction

Agro-processing comprises manufacturing activities that transform raw materials and intermediate products from agriculture, forestry and fisheries into finished goods, contributing significantly to economic growth and food security (FAO, 2017; Mazungunye, 2019). The sector encompasses the manufacturing of food, beverages, tobacco, textiles, clothing, leather, footwear, wood products, paper and paper products, rubber products and furniture products (Mazungunye, 2019). The majority of present-day industrialised countries used agro-processing as an entry point to industrialisation. The industry is important for the creation of employment and incomes, and as a strategy for pro-poor growth in rural economies; therefore, expanding agro-processing has a positive impact on human development (Mazungunye, 2019; Wilkinson & Rocha, 2009).

Agro-processing is attracting attention in Africa as it is realized that it may be a way forward for the majority of Africa's inhabitants to address poverty. Poverty is defined as a pronounced deprivation of well-being (Sharp, 2019). According to Sharp (2019), the percentage of the population in sub-Saharan Africa living at or below $1.25 a day decreased over the past few decades, from 57.6% in 1990 to 50.9% in 2005; however, the absolute number of people living in poverty has increased during that time. The poverty rates in Sub-Saharan Africa (SSA) are predicted to increase by more than 2% by 2030, delaying poverty elimination in the region by almost 5 years. Collier (2007) noted that absolute poverty, worldwide has been declining for approximately the past 25 years, yet in Africa it increased. The percentage of the extremely poor in the world fell from 40 to 18%, whereas in Sub-Saharan Africa, the numbers increased by 44% between 1981 and 2004 (Kates & Dasgupta, 2007). The COVID-19 pandemic and other emerging diseases have worsened the situation, increasing poverty among resource-poor food producers, particularly in fragile regions (Hendriks et al., 2021). Therefore, Africa's challenge is to break out of an economic stagnation that has persisted for the past three decades (Collier, 2007). In a study on how people live in a system that shows non-equilibrium dynamics, Poshiwa (2013) also concluded that it was only through industrialisation that people could meet their aspirations. It would reduce reliance on primary production and subsequently lower dependence on natural resources, thereby increasing the potential for sustainable management of those resources. Therefore, as the global population grows and environmental concerns escalate, sustainable industrialisation in agro-processing becomes critical. This paper examines how sustainable practices in agro-processing can address challenges related to resource use, waste management and environmental impact while promoting economic development.

Challenges of Agro-processing

Promoting the agro-processing industry in Africa is however, not without challenges. Poor access to infrastructure, finance and markets; limited development of entrepreneurship and managerial capacity; and a lack of an enabling regulatory environment are the most prominent and common constraints facing the continent's agro-processing sector (Woldemichael et al., 2017).

Access to Infrastructure

Sustainable Development Goal (SDG) 9 calls for states to 'build resilient infrastructure, promote inclusive and sustainable industrialisation and foster innovation through inclusive and sustainable industrialisation' (Moyo, 2017). In this context, developing a competitive agro-processing industry in Africa requires adequate physical infrastructure. Agricultural raw materials are bulky and perishable; hence, access to infrastructure, such as roads, railways, energy electricity, water and communication services at the lowest possible cost is critical for globally competitive agro-processing industries (Woldemichael et al., 2017). Post-harvest losses must be minimising for products like fruits, vegetables and dairy products through refrigeration during transportation from farms to processing plants or storage facilities. Interruptions in electricity or deficient transportation infrastructure can severely disrupt supply chains and increase production cost (Woldemichael et al., 2017). For efficient information exchange between small-scale farmers, agricultural traders, agro-processing plants and global markets; access to low-cost communication infrastructure through availability of internet and mobile services is also a critical requirement. Africa, however, faces a significant infrastructure deficit, and nurturing its agro-processing industry necessitates overcoming these bottlenecks through bold strategies and strategic investments (Woldemichael et al., 2017).

Access to Finance

Small-scale farmers, who form the majority of agricultural producers in Africa, face significant challenges in accessing finance. These farmers are constrained by low productivity, outdated technologies and limited access to modern inputs, such as fertilisers, improved and drought-resistant seeds and modern storage facilities (Woldemichael et al., 2017). A key barrier to adopting modern technologies and improving farm practices is their restricted access to financial resources. Credit is often rationed, and when available, farmers frequently lack the collateral required by banks, face high interest rates and encounter cumbersome and inefficient loan processing procedures (Woldemichael et al., 2017). Furthermore, commercial banks' reluctance in lending credit to the agricultural sector compared to other sectors, such as manufacturing and service sectors is a cause for concern. Inadequate government expenditure on agriculture compounds the challenges of achieving inclusive agricultural finance. Specifically, many financial institutions consider agribusinesses to be high-risk investments, leading to higher interest rates and stricter loan conditions, making it difficult for them to secure funding. Small-scale and medium-sized enterprises (SMEs) may not have significant assets; hence, fail to secure loans due to insufficient collateral. Furthermore, loan repayment can be unfavourable, with short repayment periods and high penalties for late repayment terms, which can strain the financial stability of agro-processing businesses. There is also a lack of incentives for private sector investment in agro-industrial processing, discouraging potential investors from entering the market. Furthermore, poor energy and water infrastructure can significantly reduce the profitability of agribusiness by increasing operational costs.

Access to Market

Lack of access to international and regional markets hinders the development of the agro-industry in Africa. Although there has been considerable progress in the liberalisation of trade in agricultural commodities since the 1980s Uruguay Round Agreement of the GATT, the agricultural sector remains one of the most protected and distorted sectors in the world (Woldemichael et al., 2017). Advanced economies continue to heavily subsidise their agricultural sectors and impose heavy tariffs on developing countries' exports of agro-processed commodities, such as food, textiles and leather products. Furthermore, high quality and traceable standards set by advanced countries, often driven by multinational retail chains, make access to the international markets difficult for medium- and small-scale agro-enterprises in Africa (Woldemichael et al., 2017; Yumkella et al., 2011). Slow regional integration at both sub-regional and continental levels also impedes industrialisation. Regional value chain processing in various sectors, for example, mineral beneficiation and agro-processing, that may depend on economies of scale cannot be implemented until regional integration takes place (Moyo, 2017). Intra-Africa trade remains limited due to inadequate infrastructure, unharmonised customs and transit procedures and other barriers (Woldemichael et al., 2017). Although efforts to promote regional integration are underway, intra-Africa trade in the agro-processing industry remains minimal.

Entrepreneurial and Managerial Skills, and Availability of Trained Personnel

Lack of entrepreneurial and managerial skills is also a key constraint that needs to be addressed. For the agro-processing industry to be competitive in the global market, technology-augmented enterprise managerial skills for efficiency and productivity are essential. While African states have significantly improved access to education services, there is a serious gap in Science, Technology, Engineering and Mathematics (STEM) education, which is critical for industrial processes (Moyo, 2017). Investment to advance skills critical for agro-processing in Africa should focus on strengthening agricultural technical and vocational education and training (TVET). Vocational training should be enhanced for professions across the entire value chain.

Facilities for Technological Innovation

Africa lacks facilities for technological innovation due to chronic under investment in agricultural research and innovation. Science, technology and innovation (STI), with adequate funding, have the potential to unlock sustainable solutions to challenges facing food systems and the agro-processing industry in Africa. Innovation in food processing and packaging is critical. Hence, it is imperative that Africa harnesses STI to promote product diversification for nutritious food, processing to extend shelf life, improved storage, food safety, extended seasonal availability and reduce post-harvest losses and food waste. The lack of investment in research and development in Africa can be addressed by partnering with international organisations and promoting strong South-South cooperation. Investment in digital infrastructure is equally important due to the complex nature of food systems, which requires transdisciplinary collaboration and inter-sectoral governance. Digitisation can enhance learning among stakeholders and across disciplines, supporting innovation and the emergence of context-appropriate technologies through transdisciplinary collaboration.

Sustainable Industrialisation

Industrialisation is a process of structural change and traditionally, the term has referred to a shift in the sectoral composition of economies starting from agriculture, moving into manufacturing and eventually into knowledge-intensive services (Gabardo et al., 2017; Kuznets, 1973; Lema et al., 2022). Industrialisation as a result is synonymous with economic development driven by manufacturing and high-value services (Szirmai et al., 2013). These activities create pathways for developing countries to grow their economies because they provide employment and linkages to other sectors of economy, and they can boost consumption and generate more foreign currency through value-added exports (Lema et al., 2022; Opuku & Yan, 2019). The theme of industrialisation has resurfaced again on the African development agenda. Concerned by the persistence of poverty, unemployment, inequality and dependence on commodity or primary production, African states appear to have re-committed themselves towards a more rapid approach for the implementation of industrialisation (Moyo, 2017). It is now argued that African countries have no option but to industrialise through manufacturing (Oyelaran-Oyeyinka & Adesina, 2020) or at least that industry needs to play an essential role in economic development, for example, with an increase in agro-industry (Lundvall & Lema, 2014). Therefore, there is an increasing agreement that economic development may involve some degree of traditional manufacturing activities, and agricultural and commodity processing, but may also involve entirely new types of industry and services (Lema et al., 2022).

Sustainable industrialisation can broadly be defined as having a dual meaning. Firstly, it is environmentally friendly industrialisation; more precisely, it gives rise to an increase in industrial activities (typically manufacturing and related services, but also other sectors e.g., agro-industry or high-value services associated with the digital economy) which do not conflict with the principles of sustainable development as they are commonly defined (Lema et al., 2022). Secondly, it is a type of industrial development that can be maintained in the long run; where industrial activities are enduring and rooted firmly in the local economy. Oyelaran-Oyeyinka (2016) contends that development is an ultimate outcome of sustainable, equitable industrialisation. In other words, it must not be industrialisation as an end in itself but rather, a means to address the social, economic and political needs of a nation. Industrialisation should thus be viewed as a strategy that empowers citizens to become independent and self-determining in all aspects of their lives (Moyo, 2017). Sustainability is also about protecting the environment. Green industrialisation is a call to ensure that Africa's industrialisation does not increase carbon emissions and destroy the environment (Moyo, 2017). This also implies contributing to social inclusiveness through the creation of jobs and incomes in local communities. In the broad sense, 'sustainable industrialisation' is industrial development that is sustainable in the environmental, social and economic senses of the term (Lema et al., 2022).

Sustainable Industrialisation in Agro-processing

The manufacturing sector in African countries is mainly dominated by agro-processing activities. These activities contribute to 52% in low income countries, 36% in middle income countries and 32% in upper middle income countries of the total value-added of manufacturing (Mazungunye, 2019). This can be even higher in agro-based countries. Athukorala (1996) points out that manufacturing is critical and likely the most important engine for long-term growth and development. The experiences of Malaysia, China and South Korea show that industrialisation led to export growth and massive increases in employment (UNIDO, 2016). Furthermore, African food systems are diverse, incorporating both traditional and modern technologies. Agriculture, including crop production, animal husbandry, fisheries and forestry and the manufacturing and their processing, can stimulate economic growth and transformation by raising rural incomes, creating jobs and increasing government revenue (Baumüller et al., 2021; Hendriks et al., 2021). Higher incomes for producers and processors can positively impact poverty reduction, food security and nutrition (Baumüller et al., 2021). The continent's major challenge remains its continued dependence on commodity or primary product production, preventing it from fully benefiting—from its rich natural resource base: oil, gas, minerals and agricultural land (Moyo, 2017). Thus, agro-processing has a vital role in contributing to the output of the economy (Mazungunye, 2019). Agro-processed products play a vital role in boosting economic output (Wilkinson & Rocha, 2009). Processed products hold higher value than raw agricultural produce, enabling value addition to exports. Exports are crucial for developing countries, as they create external demand and stimulate increased production (Fukase & Martin, 2017; Mazungunye, 2019).

Principles of Sustainable Industrialisation in Agro-processing

Efficiency and Resource Management

Sustainable industrialisation emphasises improving efficiency and optimising resource use. Improving the efficiency of production systems is necessary due to the constraints on land and resource availability, and the relatively small land plots common in most of Africa (Lowder et al., 2016). Improving production efficiency is necessary to meet the growing demand for food (including animal-sourced foods) and is also an environmental imperative. The Food Systems Summit calls for a shift to nature-positive production systems that meet the fundamental human right to healthy food while operating within planetary boundaries that define the limits of natural resources available for sustainable exploitation. In agro-processing, improving efficiency and optimising resource use can be achieved by adopting technologies and practices that enhance productivity while minimising waste and resource consumption. Key strategies include using energy-efficient machinery, optimising water use and implementing waste reduction techniques (FAO, 2021).

Environmental Impact Mitigation

Mitigating environmental impact is a cornerstone of sustainable industrialisation. Sustainability is about protecting the environment; thus, green industrialisation is a call to ensure that Africa's industrialisation does not increase carbon emissions or destroy the environment (Moyo, 2017). This involves the use of 'green technologies', technologies that have the least possible impact on the environment, achieved through an industrialisation process that is 'green' in both process and outcome (UNIDO, 2009; World Bank, 2012). Therefore, agro-processing facilities must implement measures to reduce emissions, manage waste effectively and minimise pollution. Innovations such as waste-to-energy systems and closed-loop water recycling are crucial for reducing the ecological footprint of agro-processing operations (Dube et al., 2020).

Economic and Social Benefits

Sustainable industrialisation not only addresses environmental concerns, but also provides economic and social benefits. Most agro-processing industries are, by nature, close to sources of raw materials and to where main agricultural activities take place (Woldemichael et al., 2017). Thus, the sector helps to increase off-farm employment opportunities in rural areas and to alleviate poverty. Additionally, due to its forward and backward linkages, the agro-processing sector encourages a range of supplementary services and supporting activities in the secondary and tertiary sectors, including informal and small-scale enterprises. Youth and women are particularly well-positioned to benefit from such employment opportunities (Woldemichael et al., 2017). Therefore, by adopting sustainable practices, agro-processing industries can enhance their competitiveness, create job opportunities and improve the livelihoods of local communities. Social sustainability involves ensuring fair labour practices and contributing to community development (World Bank, 2020).

Technological Innovations in Sustainable Agro-processing

Energy Efficiency and Renewable Energy

Energy efficiency and the use of renewable energy sources are vital for sustainable agro-processing. Technologies such as energy-efficient machines, solar panels and biogas systems help reduce energy consumption and reliance on fossil fuels. For example, the integration of solar energy in agro-processing plants has proven effective in lowering operational costs and reducing carbon emissions (Bhattacharyya, 2018). Africa has a substantial and underutilised potential for renewable energy production, particularly in solar energy harvesting. In many locations across the continent, the use of renewable energy sources, often integrated into decentralised local grids, can be less costly than fossil-fuel-based electricity generation. Renewable energy development also has substantial employment generation potential. In fact, many governments across Africa are identifying the renewable energy sector as a major area for action for creating 'green jobs', which could generate year-round employment for young people engaged in the agricultural sector (Malabo Montpellier Panel, 2019). Investing in renewable energy resources in Africa will also have important implications for climate change mitigation and sustainable land management (Mirzabaev et al., 2019). The climate change mitigation potential comes not only from a reduction in greenhouse gas emissions due to reduced use of fossil fuels, but also through increased carbon sequestration achieved by reducing deforestation for charcoal and fuelwood production.

Waste Management and Circular Economy

Effective waste management and the implementation of circular economy principles are essential for sustainability. Innovations such as anaerobic digestion, composting and recycling systems help manage waste and convert it into valuable resources. The adoption of circular economy practices ensures that waste is minimised and materials are reused, contributing to a more sustainable production cycle (Nascimento et al., 2019). By integrating these waste management strategies and circular economy principles, agro-processing industries can reduce their environmental impact, improve sustainability and create economic opportunities.

Smart Technologies and Automation

Smart technologies and automation enhance efficiency and sustainability in agro-processing. Africa is at the beginning of the 'Fourth Industrial Revolution' (4IR) or

'Smart' industrialisation (Moyo, 2017). Currently, a very small number of industrial units produce medium to high technology manufactured products. Instead, their manufacturing largely consists of primary commodities, with a small proportion of light-technology goods (Moyo, 2017). The 4IR involves the digitisation of industrial processes, which include the use of artificial intelligence, robotics, nanotechnology, biotechnology and other related technologies (Moyo, 2017). The technologies of the 4IR have the potential to improve productivity and reduce energy and resource consumption, making them both efficient and environmentally friendly. Sustainable industrialisation involves adding value to the natural resources on the continent while minimizing damage to the environment. Digital technologies such as IoT sensors, automation systems, artificial intelligence (AI) and data analytics enable real-time monitoring and optimisation of production processes. These innovations lead to improved product quality and safety, reduced waste and increased operational efficiency (Pereira et al., 2021). The agro-processing industry has the potential to undergo a revolutionary transformation by adopting such technologies.

Strategic Recommendations for Promoting Sustainable Industrialisation in Agro-processing

Policy Frameworks and Incentives

African countries need to integrate the industrialisation agenda into their national visions, development plans and strategic frameworks. Governments should develop supportive policy frameworks and incentives to encourage sustainable practices in agro-processing. This includes providing financial support for renewable energy projects, implementing regulations that promote waste reduction and offering tax incentives for adopting energy-efficient technologies (OECD, 2019).

Investment in Rural and Agricultural Finance

Investment in the establishment and expansion of dedicated agricultural finance institutions is crucial to accelerate capitalisation of the agricultural production sector and to improve risk management. Interventions are needed to strengthen the governance and management of agricultural development banks. Additionally, rural cooperatives and rural credit and savings associations should be linked to agricultural banks. Micro-finance systems should be further expanded as part of the institutional landscape for the productive and rural service sectors.

Public–Private Partnerships

Public–private partnerships (PPS) can play a crucial role in advancing sustainable agro-processing. Collaborations between government agencies, private companies, entrepreneurs research institutions can facilitate the development and implementation of innovative technologies and practices. Such partnerships can also help overcome financial and technical barriers to sustainability (UNIDO, 2022).

Research and Development

Agricultural research is one of the most effective investments to unlock potential. For example, it is estimated that every US$ spent on national agricultural R&D generates average returns of US$ 3 (Fuglie & Rada, 2013). Investing in research and development is essential for advancing sustainable agro-processing technologies. Funding for research projects and technology development can lead to new solutions for improving efficiency, reducing waste and minimising environmental impact. Collaboration between academia, industry and government can drive innovation in this sector (FAO, 2021). The benefits of agricultural R&D could be amplified by improving linkages between formal agricultural research institutes and the private sector and informal structures among producers, processors and civil society groups to ensure more effective and participatory research processes and the adoption of innovations. Efforts should be made to improve linkages between agricultural research and extension providers to promote the widespread dissemination of promising innovations.

Capacity Building and Training

Capacity building and training programs are vital for equipping the workforce with the skills needed to implement and manage sustainable practices. It is critical for Africa to prioritise investment in the people who manage African agriculture and the food system, especially agro-processing. Facilities should be established for agricultural, technical and vocational education and training (TVET), drawing on successful examples from the continent and experiences from other countries. Educational initiatives should focus on emerging technologies, sustainability practices and efficient resource management to ensure a knowledgeable and skilled workforce (World Bank, 2020). Training programs should be organised and technical assistance provided in the basic principles of processing, hygiene and sanitation, quality assurance and standards, product development, food microbiology, labelling and packaging for human resource development and technology transfer in agro-processing.

Infrastructure Development

SDG 9 calls for states to 'build resilient infrastructure, promote inclusive and sustainable industrialisation and foster innovation through 'inclusive and sustainable industrialisation'. In order to achieve this, Africa needs to develop its infrastructure, particularly roads, rail, ports and air transport, to enhance competitiveness through cutting down the cost of doing business. Furthermore, the African continent needs to prioritise investments in electricity generation, water services provision, telecommunications and ICT (Moyo, 2017). Investments in infrastructure for mobile connectivity, in particular, are a high-payoff priority across Africa as a prerequisite for digital tools to be widely and effectively used in the food and agriculture sector. It is also necessary to have adequate refrigeration facilities for the storage and transportation of perishable agricultural produce.

Conclusion

Many African countries remain largely agriculture-based, and numerous value chains are still underdeveloped. Sustainable industrialisation in agro-processing could unlock significant opportunities for economic and development across the continent. The development of Africa's food systems and agro-processing sectors holds the potential to foster healthier, more sustainable and equitable food systems, particularly when supported by advanced technologies and research.

Sustainable industrialisation in agro-processing is essential for addressing environmental challenges while simultaneously promoting economic growth and social progress. By leveraging technological innovations, adopting best practices and implementing supportive policies, the agro-processing sector can achieve significant improvements in efficiency, resource management and environmental stewardship. The strategic recommendations outlined in this paper provide a roadmap for advancing sustainable practices and ensuring a resilient and prosperous future for agro-processing in Africa.

References

Athukorala, P., & Menon, J. (1996). Foreign investment and industrialisation in Malaysia: Exports, employment and spill overs. *Asian Economic Journal,* 29–44.

Baumüller, K., Admassie, A., Hendriks, S., Tadesse, G., & von Braun, J. (Eds.). (2021). *From potentials to reality: Transforming Africa's food production—Investment and policy priorities for sufficient, nutritious and sustainable food supplies.* Peter Lang Publ. Forthcoming an earlier draft is available at https://www.zef.de/fileadmin/downloads/ZEF_Akademiya2063.pdf

Bhattacharyya, S. C. (2018). *Energy for sustainable development: A study on energy efficiency in agro-processing industries.* Springer.

Collier, P. (2007). Poverty reduction in Africa. *PNAS, 104*(43), 16763–16768.

Dube, T., Mbohwa, C., & Tsoka, S. (2020). *Waste management and circular economy in agro-processing*. Elsevier.

FAO. (2017). *The state of food and agriculture*. FAO.

FAO (2021) Policies and incentives for sustainable agro-processing. *Food and Agriculture Organization of the United Nations.*

Fukase, E., & Martin, W. (2017). *Agro-processing and horticultural exports from Africa*. Discussion Paper 01690. IFPRI.

Fuglie, K & Rada, N (2013). Resources, policies and agricultural productivity in Sub-Saharan Africa. *USDA-ERS Economic Research Report Number*, 145.

Hendriks, S. L., Bekele, E., Chaibi, T., Hassan, M., Miano, D. W., & Muyonga, J. H. (2021). *The role of science, technology, and innovation for transforming food systems in Africa*. Food Systems Summit Brief Prepared by Research Partners of the Scientific Group for the Food Systems Summit

Gabardo, F. A., Pereima, J. B., & Einloft, P. (2017). The incorporation of structural change into growth theory: A historical appraisal. *Economia, 18*(3), 392–410. https://doi.org/10.1016/j.econ.2017.05.003

Kates, R. W., & Dasgupta, P. (2007). African poverty: Agrand challenge for sustainability science. *Proceedings of the National Academy of Sciences of the United States of America, 104*(43), 16747–16750.

Kuznets, S. (1973). Modern economic growth: Findings and reflections. *America Economic Review, 63*, 247–258.

Lema, R., Andersen, M. H., Hanlin, R., & Nzila, C. (2022). *Building innovation capabilities for sustainable industrialisation renewable electrification in developing economies*. Routledge.

Lowder, S., Skoet, J., & Raney, T. (2016). The number, size, and distribution of farms, smallholder farms, and family farms worldwide. *World Development, 87*, 16–29. https://doi.org/10.1016/j.worlddev.2015.10.041

Lundvall, B., & Lema, R. (2014). Growth and structural change in Africa: Development strategies for the learning economy. *African Journal of Science, Technology, Innovation and Development, 6*, 455–466. https://doi.org/10.1080/20421338.2014.979660

Malabo Montpellier Panel (2019). *Energized: Policy innovations to power the transformation of Africa's agriculture and food system*. Malabo Montpellier Panel.

Mazungunye, P. P. (2019). *Industrialisation for economic transformation. Economy-wide impacts of agro-processing development in Tanzania*. SA-TIED Working Paper #42.

Mirzabaev, A., Wu, J., Evans, J., Garcia-Oliva, F., & Hussein, I., et al. (2019). Desertification. In: P. R. Shukla et al. (Eds.), *Climate change and land: An IPCC special report on climate change, desertification, land degradation, sustainable land management, food security, and greenhouse gas fluxes in terrestrial ecosystems* (pp. 249–343). World Meteorological Organization and United Nations Environment Programme.

Moyo, T. (2017). Promoting inclusive and sustainable industrialisation in Africa: A review of progress, challenges and prospects. In *The 2nd Annual International conference on public administration and development alternatives*, 26–28 July 2017, Tlotlo Hotel, Gaborone, Botswana.

Nascimento, A. S., Silva, M. L., & Rodrigues, C. (2019). *Waste-to-energy technologies in agro-processing: A Brazilian perspective*. Springer.

OECD. (2019). *Research and development in agro-processing: Innovations and impact*. Organisation for Economic Co-operation and Development.

Opoku, E. E. O., & Yan, I. K. M. (2019). Industrialisation as driver of sustainable economic growth in Africa. *The Journal of International Trade & Economic Development, 28*(1), 30–56. https://doi.org/10.1080/09638199.2018.1483416

Oyelaran-Oyeyinka, B., & Gehl Sampath, P. (2016). *Sustainable industrialisation in Africa: Towards a new development agenda*. Palgrave, Macmillan.

Oyelaran-Oyeyinka, B., & Adesina, A. A. (2020). *Structural transformation in African development: Industrialisation and urbanization pathway*. Anthem Press.

Pereira, C., Sorribas, M., & Martín, M. (2021). *Smart technologies in agro-processing: Enhancing efficiency and sustainability*. Springer.
Poshiwa, X. (2013). *African wildlife and people: Finding solutions where equilibrium models fail* [Ph.D. Thesis, Wageningen University and Research].
Sharpe, R. (2019). Poverty in global perspective. In N. A. Naples (Ed.), *The Wiley Blackwell encyclopedia of gender and sexuality studies* (1st edn.). Women's Institute for Science, Equity and Race. Bucknell University, USA KENDALSWANSON Duke University, U.S. Wiley. https://doi.org/10.1002/9781118663219.wbegss279
Szirmai, A., Naudé, W., & Alcorta, L. (2013). *Pathways to industrialisation in the twenty-first century: New challenges and emerging paradigms*. Oxford University Press Online.
United Nations Industrial Development Organization (UNIDO). (2009). *A greener footprint for industry: Opportunities and challenges of sustainable industrial development*. UNIDO.
United Nations Industrial Development Organisation (UNIDO). (2016). *Industrialisation in Africa and least developed countries. Boosting growth, creating jobs, promoting inclusiveness and sustainability*. A Report to the G20 Development Working Group by UNIDO.
UNIDO. (2022). *Training and capacity building for sustainable agro-processing*. United Nations Industrial Development Organization.
Wilkinson, J., & Rocha, R. (2009). Agro-industry trends, patterns and development impacts. In C. A. da Silva, D. Baker, A. W. Shepherd, C. Jenane, & S. Miranda-da-Cruz (Eds.), *Agro industries for development*. CABI and FAO.
Woldemichael, A., Adeleke, S., Adamon, M., Anthony, S., & Abebe, S. (2017). Transforming Africa's agriculture through agro-industrialisation. *Africa Economic Brief, 8*(7).
World Bank. (2012). *World development report 2012: Gender equality and development*. The World Bank.
World Bank. (2020). *Public-private partnerships in agro-processing: Opportunities and challenges*. World Bank Group.
Yumkella, K., Kormawa, P., Roepstorff, T., & Hawkins, A. (2011). *Agribusiness for Africa's prosperity*. United Nations Industrial Development Organization (UNIDO).

Chapter 11
The Impact of Infrastructure Development in the Deployment of Internet-Based Technological Innovations in Industrial Clusters

Olomu Michael Oluwaseun, Binuyo Gbonjubola Oluwafunmilayo, Oyebisi Timothy Oyedepo, and Onifade Olabanji John

Abstract The deployment of technological innovation is crucial to the realization of economic growth and development. However, in today's market-based globally competitive environment, industrial clusters' firms in developing economies face stiff competition and increasing demands for high quality products and services. These demands are characterized by fast response times, reliable deliveries and innovative product functionalities. Despite the rapid pace of infrastructural development globally in recent years especially in the industrial sectors, the opportunities and challenges associated with industrial clusters in developing countries have grown increasingly complex overtime. As a result, the next frontier lies in understanding how technological changes and innovations impact industrial clusters, infrastructural development and overall competitiveness. The study examined the extent of infrastructure development in the deployment of Internet-based technological innovations in the industrial clusters. The results reveal that the most common infrastructure development activities among the industrial clusters' firms entail the acquisition of networking equipment and technological infrastructure. However, the leasing or rentals of equipment, as well as external linkages, were among the least common

O. M. Oluwaseun (✉)
Department of General Studies, Federal University of Technology and Environmental Sciences (FUTES), Iyin-Ekiti, Nigeria
e-mail: michael.olomu@futes.edu.ng

B. G. Oluwafunmilayo · O. T. Oyedepo
African Institute for Science Policy and Innovation (AISPI), Obafemi Awolowo University (O.A.U), Ile-Ife, Nigeria
e-mail: gobinuyo@oauife.edu.ng

O. T. Oyedepo
e-mail: toyebisi@oauife.edu.ng

O. O. John
Science Policy and Innovation Studies (SPIS) Department, National Centre for Technology Management (NACETEM), Ile-Ife, Nigeria

X. Poshiwa et al. (eds.), *Technological Innovations for Sustainable Industrialization*,
https://doi.org/10.1007/978-981-96-7995-9_11

activities carried out by the firms. Additionally, it was observed that there was a low level of investment and expenditure on infrastructure development by the industrial clusters' firms. Thus, the study suggests the need to strengthen infrastructural development, as well as industrialization and sustainable development of the industrial clusters, as this would increase and sustain firms' performance, boost productivity and competitiveness, contribute to economic growth and ultimately generate employment opportunities. Similarly, the right investment in Internet network infrastructure can bring down consumer prices and improve affordability. Therefore, the availability of infrastructural and institutional support could play a critical role in the development of industrial clusters, while innovation and infrastructure development are crucial in creating a competitive advantage for firms within industrial clusters.

Keywords Internet · Technological innovations · Fourth industrial revolution (4IR) · Industrial firms · Industrial clusters · Infrastructure · Development · Nigeria

Introduction

The adoption of technological innovation is crucial to the realization of economic growth and development, notably driven by numerous factors and conditions (Venkatesh et al., 2012). In the present globalized economy, industrial firms face stiff competition and growing demands for high quality products and services in the market. Moreover, global business disruptions and uncertainties underscore the need for increased and effective infrastructure development and sustainable industrialization, which will necessitate enhanced outsourcing activities, transformations in value chains and distribution channels.

Notably, leveraging technological innovations has led to the integration of different economic components in a remarkable way, providing new means of communication capabilities and facilitating information exchanges within firms, among customers and across clusters in the global supply chain (Somuyiwa & Oyesiku, 2010).

Thus, infrastructural development and advancement are crucial in creating a competitive advantage for firms within industrial clusters, since such firms tend to enjoy greater access to information, infrastructure, suppliers and distributions networks than other firms (Pandit et al., 2018).

The world is at the dawn of the Fourth Industrial Revolution (4IR), as processing, management and storage capacities are expanding exponentially. Thus, knowledge is becoming available to more people than ever before in human history (World Economic Forum, 2016).

Moreover, the future holds greater potential for human advancement as the full impacts of infrastructure development and emerging technological changes such as artificial intelligence (AI), the Internet of Things (IoT), smart automation, energy storage, 3-D printing, advanced wireless technologies and dynamic computing continue to evolve.

The accelerating pace of infrastructure development, widespread disruption across all major industrial segments and the transformation of entire systems of production and management are what distinguish these rapid advancements from earlier 'industrial revolutions' (World Economic Forum, 2016).

For this new revolution to be fully actualized, the efficient deployment of innovations for infrastructure development is crucial, as the future prospects of industrial sectors will rely more than ever on whether they strategically adopt technological innovations or not.

In recent years, the Internet has expanded exponentially across industrial sectors, becoming one of the most crucial technologies of the twenty-first century. In fact, it forms the foundation of the 'Fourth Industrial Revolution' (4IR) and digital transformation of society and business (Lee et al., 2014). The impact of Internet technology on the sustainable industrialization is undeniable. This is because technological innovation and capabilities are increasingly becoming a critical component and driving force industrial growth, which is essential for any sustainable industrial development and long term progress.

The introduction of Internet technology has enhanced productivity among industrial firms and has gradually helped them to manage and sustain both intra- and inter managerial operations. Moreover, the global advancement of Internet technology has positioned numerous industrial firms worldwide to adapt their business and organizational strategies in search of new ways to improve productivity and profitability (Lieven et al., 2011).

Hence, the widespread deployment of Internet-based technological innovations which are operations, services or activities performed using the communication infrastructure of the Internet or Internet-enabled devices has also transformed the ways in which industrial communication is conducted.

Internet technology has also transformed the procedures, systems and processes of service delivery and consequently shapes the ways in which industrial suppliers, customers and other external entities interact. Meanwhile, the successful implementation of Internet-based technology across the industrial sector has a substantial effect on technological infrastructure development and usage behaviours in emerging countries (Zhang, 2018).Internet-based technological innovations are gradually advancing towards the managerial and strategic adoption of AI, autonomous machine learning, data analytics and robotics within industrial clusters, and this has reshaped the core infrastructure and operations, leading to a digital enterprise transformation.

Internet-based technological innovations seeks to integrate complex processes and business functions, such as management, marketing strategy, production, manufacturing, finance, accounting, human resources, customer relationship management, logistics and many other industrial operations. The advent of the Fourth Industrial Revolution (4IR) has provided alternative paths to industrialization and economic development for African countries, and with the infrastructural transformation now well underway; another conceptual shift is required to understand the evolving role of infrastructure development for sustainable industrialization.

It is evident that infrastructural breakthroughs in the global markets have a spillover effect on the structural settings of African economies' industrialization, as rapid technological changes have fragmented production across borders, but some African countries remain marginalized.

However, sustainable industrialization is integral for Africa to achieve economic growth, while leveraging on technological innovation and infrastructure development as key enablers.

The United Nations Sustainable Development Goal-9 (SDG 9) as well as the Africa's Agenda 2063, assert the role of infrastructure and industrialization in economic transformation, job creation, increased productivity and achieving sustainable development.

Notably, significant progress has been made as several efforts aim to bridge the digital divide, but the challenges remain enormous and complex, necessitating collaborative efforts from industrial sector players in Africa to assess and overcome the four major barriers to the inclusion of Internet-based technological innovations: infrastructure affordability, skill acquisition, awareness and cultural acceptance.

These challenges have strategically led to the formation of the several initiatives and programmes aimed at boosting infrastructural development in Africa, such as the Internet for All Initiative, the Policy and Regulation Initiatives for Digital Africa (PRIDA), the Continental ICT Strategy for Africa (CISA), the Harmonization of ICT Policies in sub-Saharan Africa (HIPSSA), the Africa Connect Project, among others.

Similarly, the Nigerian government has introduced various initiatives, primarily to support industrial clusters and focus on addressing the challenges of sustainable industrialization and infrastructure development (Ugoani & Ibeenwo, 2015). Industrial clusters have triggered continuous economic growth in the sub-Saharan Africa, many through industrial infrastructure, technology and financial development (Bank of Industry, 2018).

With this understanding, the Nigerian government has made efforts to recognize, promote and channel support to industrial clusters through proactive and coordinated efforts through financial actors and development partners, in response to the rising global market and competitiveness.

The gaps in innovation, technology and skills among developed, developing and least developed economies persistently and significantly widened due to the vast asymmetry in access to new knowledge, capabilities, information, emerging technologies and modern infrastructure, as well as limited access to technological learning and innovation channels. However, the deployment of Internet-based technological innovations supported firms in becoming globally competitive, but various challenges hinder Nigerian industrial firms effectively acquiring and utilizing innovative and modern infrastructure for sustainable industrialization (Olomu et al., 2023).

Furthermore, owing to far-reaching globalization, industrialization, technological change, convergence and competition, innovation has resulted in an infrastructure development spiral. Within this spiral, an increasing number of modern, Internet-based technological infrastructure are launched in ever shorter time intervals, reducing product life-cycles but boosting user adoption (Lieven et al., 2011; Zhang, 2018).

Despite the rapid pace of infrastructural development in recent years, especially in the industrial sector, the opportunities and challenges associated with industrial clusters have become more complex overtime. The next frontier is understanding on how technological changes and innovation impact industrial cluster infrastructure development and competitiveness and evaluating their usefulness for policymaking. Equally important is determining how industrialization and infrastructural development can provide guidance to harnessing technological changes and innovation, which has become imperative. Industrial clusters infrastructure development has led to more diverse paths of structural transformation resulting in a phenomenon characterized by the pervasive and quickening pace of technological change and globalization (Naudé and Nagler, 2015). Moreover, industrial clusters across the globe now extensively deploy Internet innovations in diverse ways to maintain their market shares and guarantee profitability in the face of growing intense competitiveness and product homogenization (Olomu et al., 2023). This study examines the extent of infrastructure development in the adoption and the diffusion of Internet-based technological innovations in the Nigerian industrial clusters.

The Fourth Industrial Revolution

The 'Fourth Industrial Revolution' (4IR) is characterized by accelerated digital transformation, notably from ICT development. Recently, it has been recognized that the 'Fifth-Generation (5G)' digital broadband cellular network technology standard, which officially emerged globally in 2019 and provides connectivity to most recent smart phones, is pivotal to the fourth industrial revolution (Roblek et al., 2020). Similarly, the 4IR represents a new level of human advancement in ICT technology, as the full impacts of infrastructure development and emerging technological changes such as the artificial intelligence (AI), the Internet of Things (IoT), smart automation, energy storage, 3-D printing, advanced wireless technologies and dynamic computing are still evolving (Fig. 11.1). The Internet has spread exponentially across economic sectors globally and it has become one of the most crucial technologies of the twenty-first century. In fact, it forms the foundation of the 4IR and digital transformation of society and business (Lee et al., 2014). The four major policy goals that drive the revolution are: (a) the world's first 5G commercialization and nationwide 5G coverage by 2022, (b) increasing IoT connectivity, (c) implementing smart and reliable network based on software and AI and (d) deploying nationwide high-speed Internet access (Hong et al., 2021).

For instance, in May 2013, the Korean Government launched the 5G Forum to initiate formal discussions on the national 5G strategies and readiness, drawing members from public institutions and the private sector.

With the establishment of the Presidential Committee on the 4IR (November 2017), the government declared three keys to the 4IR; Data, Network and AI (DNA) (Government of the Republic of Korea, 2020). Since then, the government has significantly expanded its support for DNA, announcing various policies including the I-Korea 4.0 Plan (2017); Hyper-Connected Intelligent Network Deployment Strategy (2017); AI R&D Strategy (2018); Data Industry Activation Strategy (2018); System

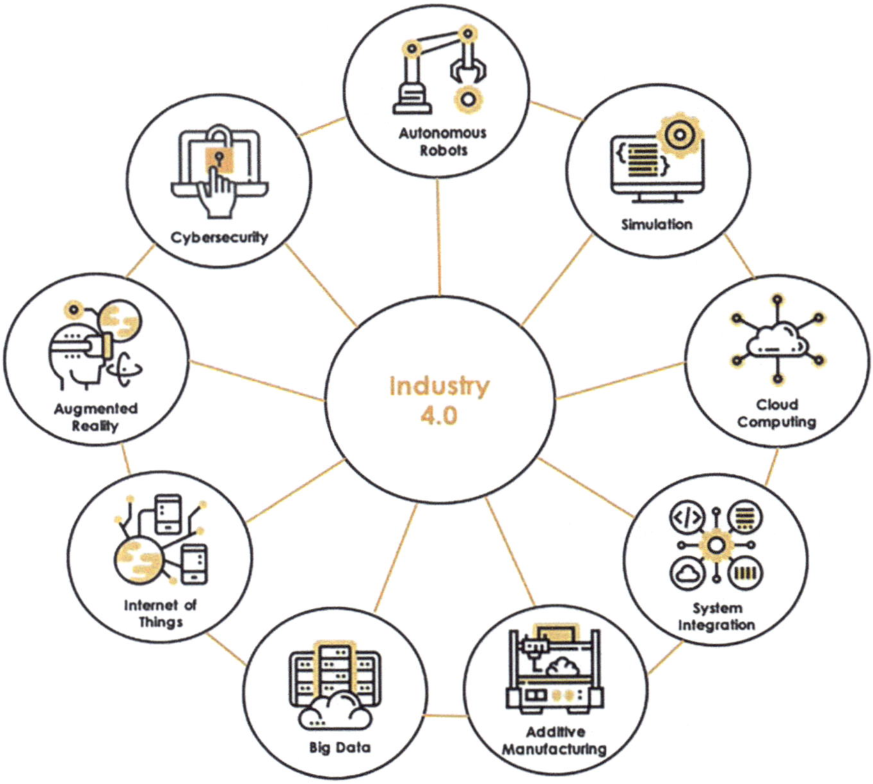

Fig. 11.1 Schematic representation of industry 4.0

Semiconductor Strategy (2019); 5G+ Strategy (2019); Manufacturing Renaissance Strategy (2019); and the Korean New Deal (2020).

Since the government announced the I-Korea 4.0 Plan in 2017 to promote the 4IR centred on enhancing people's lives based on ICT technology, it has clearly presented its vision and roadmap to introduce 5G. Following the launch of the I-Korea 4.0 Plan, the government announced the Hyper-Connected Intelligent Network Deployment Strategy (2017) to prepare for the 4IR (Fig. 11.2).

Infrastructure and Industrial Clusters Development in Nigeria

Since the growth of a country is largely dependent on industrial clusters and infrastructure development in the present market-based global competitive environment (Azam, 2015). The rapid expansion of ICT infrastructure and its spread to almost

Fig. 11.2 Infographic keys to the fourth industrial revolution. Adapted from the Government of Republic of Korea (2020)

every sphere of economic activity in Nigeria has initiated the process of encouraging industrial clusters to make the most cost-effective use of the available infrastructure in production and operations (Osarenkhoe and Fjellström, 2017). Thus, how Nigerian industrial clusters can fully tap into these technological opportunities and infrastructure to produce enhanced performance for economic development is considered critical.

The Nigeria Industrial Revolution Plan (NIRP), which was an essential part of the 2017 Economic Recovery and Growth Plan (ERGP) policy document of the Nigerian government, was formulated to facilitate the development of industrial clusters, cities and parks to ensure the availability of infrastructure within these industrial zones. The approach was to refocus the ERGP in order to accelerate its implementation strategy and safeguard economic growth, rapid industrialization and the survival of industrial enterprises in Nigeria.

Also, the Nigeria ICT Innovation and Entrepreneurship Vision (NIIEV), launched in April 2019 by the National Information Technology Development Agency (NITDA), was established to unlock and foster the growth of the technology ecosystem and also provide support systems to connect, educate and enable digital innovation capable of transforming and accelerating industrial sector and other sectors of the economy. Since 2001, NITDA has regulated the IT sector with a mandate to make policy recommendations to support technology development in Nigeria. To ensure that NIIEV is properly implemented across the targeted sectors of the Nigerian economy, NITDA established the Office for ICT Innovation and Entrepreneurship (OIIE) to drive innovation and entrepreneurship mandate of the Agency. The mandates for the establishment of OIIE are to develop, establish, support and incentivize ICT innovation hubs, information technology parks and community enterprise hubs across Nigeria; and to collaborate with private sector, development partners and MDAs at both federal and state levels on policies and programmes that support ICT entrepreneurship and commercialization. More importantly, the main target of the policy document is to ensure that 95% of Nigerian population significantly have increased access to ICT, to provide universal and affordable broadband Internet access and to ensure that ICT will contribute 25% to the national GDP by year 2025. The first vision of the document is to 'incentivize telecommunication infrastructure sharing to achieve the economies of scale necessary for universal affordable access', and to achieve this, the Nigerian government will support private sector providers in sharing their infrastructure.

In sub-Saharan Africa, many countries are striving to enhance the deployment of Internet, technological infrastructure and capabilities which are crucial to improving services and closing the widening digital divide. Presently, the total length of optical fibre throughout Africa is more than 4,850 km, while 4G wireless coverage stands at 63%. Similarly, in terms of terrestrial satellites, thirteen (13) African countries have launched 48 satellites. Although this infrastructure appears impressive, in reality, Internet penetration remains quite low compared with other continents of the world.

According to the Mobile Connectivity Index in 2019, Nigeria was the most improved country in sub-Saharan Africa and the seventh most improved globally, raising its score from 45 to 49 within a single calendar year. The growth was primarily driven by a range of improvements across the infrastructure, affordability and content and services enablers.

Also in 2019, 3G coverage stretched to 78%, with 20 million users covered during the same year. Also, since the emergence of 4G network in 2016, communication operators in Nigeria have achieved 45% 4G coverage. These breakthroughs in coverage in particularly 3G have been largely driven by the diffusion of UMTS-900 (operators redeploying part of their 2G spectrum and infrastructure in 900 MHz to diffuse 3G). This has also allowed operators to introduce and deliver 3G in a more cost-efficient way.

Similarly, as part of the Facebook's initiative to support sustainable development through investment in broadband availability and affordability in Nigeria, it has invested in Open Transport Networks (OTNs) to deploy fibre infrastructure for backhaul (ITU, 2020). The project actively involved local operators to introduce 3G/4G or deploy new network technologies, and it enabled an expansion of 3G/4G coverage to over 4 million Nigerians. The project is further envisaged to spur rapid economic development by promoting the quality and affordability of services and delivery by connecting more citizens. Thus, improved 4G coverage in Nigeria has resulted in increased network download and upload speeds and lower latencies, which underscores the need to constantly adapt regulatory framework and justify licenses retained by local operators as technology neutral.

In addition, there is significant and growing investment in edge network infrastructure in Nigeria such as Points of Presence (PoPs) and local co-located caches by the Nigerian government, aimed at promoting economic development by allowing operators to access content at locations closer to their own networks. This initiative has improved connectivity and reduced transit costs for operators, while enhancing user experience in Nigeria. The result has increased supply of bandwidth and lower operational costs for ISPs, leading to lower prices for end users.

In terms of content and Internet services in Nigeria, the Online Index score for e-Government improved significantly from 31 to 52% between 2014 and 2019. With this laudable development, Nigeria has successfully established a virtuous circle of expanding and bridging more coverage gaps, while promoting more affordable Internet-based devices and data plans. This has encouraged operators' investment and enabled the rollout of more coverage as adoption rate increases.

Furthermore, ISPs in Nigeria are making several concerted efforts to make services more affordable and accessible by easing data limits, upgrading speeds, ensuring zero-rating traffic to specific webpages and offering special low-cost voice and data packages to users. Additionally, network operators and ISPs are increasingly improving capacity of existing networks, infrastructure and technologies, while adopting a range of other emerging technologies for enhanced service deliver to encourage these service providers, the Nigerian Government is relaxing various regulatory requirements and recommendations through emergency spectrum licensing allocation, as well as the removal and reduction of charges for network and spectrum usage.

These measures would then promote zero-rating traffic, information sharing on the network, granting access for VoIP, providing subsidies for users and the acquisition of SIM (Subscriber Identification Module) cards and gadgets and providing financial assistance to service providers, network operators and broadband providers.

During the COVID-19 pandemic, many network providers and content providers in Nigeria have systematically taken steps to allow greater access to accurate and timely health information as many provide free access to their webpages, while some give free data to their customers daily. Also, there are increasing plans underway to provide broadband connectivity access to forcibly displaced persons at their hosting communities in Nigeria.

The industrial sector is focused on a drive towards continual infrastructure development; thus, sustainable infrastructure becomes a key enabler of economic and social development, as well as sustainable industrialization and environmental development.

Consequently, quality infrastructure development enhances productivity and competitiveness, contributing to economic growth and employment generation, as well as facilitating international trade.

Furthermore, infrastructure is vital for a country's accelerated economic growth and poverty reduction. For an economy to integrate with other economies across the world, appropriate infrastructure and their efficient operation are inevitable.

Some economic activities are clustered in Nigeria to create a robust and controlled environment for industrialization to flourish especially in the face of severely inadequate infrastructure and resources, particularly regarding power generation.

Traditionally, this development has taken the form of industrial estates or parks, which allow for prioritization of infrastructural facilities and give firms a competitive advantage, while offering access to skilled labour, technology and raw materials (Chete et al., 2013).

The Nigerian Industrial Sector and Fast-Moving Consumer Goods Industry

The Nigerian industrial sector comprises of manufacturing, utilities, mining, etc. and accounts for about 6% of economic activities in 2011 (Chete et al., 2013). Over the years, the contribution of the sector to the Nigeria's GDP has increased tremendously. The economic development of a country is principally dependent on industrial sector and firms in the present market-based global competitive environment. The Nigerian industrial sector is very dynamic and accounts for averagely 25.7% of the GDP, employing 12% of the workforce annually.

The Nigerian industrial sector is specifically known for production of consumer and capital goods. The consumer goods segment of the industrial sector comprises firms that are involved in the production and manufacturing of finished goods.

Generally, these products and services are classified for personal use, specifically produced for the mass market and encompass goods that are consumed rather than used in the manufacturing of other goods, including both durable and non-durable consumables.

Also in this sector are manufacturers of automobiles/auto parts, household durable goods, textiles and apparel, as well as manufacturers' food, beverages and tobacco products. Moreover, the capital goods segment of the industrial sector comprises firms that are primarily engaged in the production and distribution of capital goods, including aerospace and defence, electrical equipment, industrial machinery, packaging products, engineering and building products for industrial and consumer products. Primarily, their activities are dominated by the production of goods for commercial use. The fast-moving consumer goods (FMCG) industry represents one of the biggest industries globally and is also known as the consumer-packaged goods (CPG) industry.

It is primarily characterized as firms that produce and supply low-cost products that are usually in high demand, such as food, beverages, household cleaning utensils and personal hygiene products. The term 'fast-moving' originated from the fact that production typically has a short shelf life and consists of primarily non-durable goods. Generally, the firms operating in the FMCG industry largely enjoy a more strategic and highly innovative, high margin, high volume game as a result of a strong concentration on pooling top line sales. Their products are also nearly identical, leading to high level of price wars and competition in the industry.

Therefore, to improve profitability and competitiveness, firms deploy innovative marketing strategies and other technology driven techniques to gain and retain customer loyalty to their products. However, in Africa, where the level of poverty is still relatively high, FMCG has the significant potential to expand, as food and other necessities usually consume majority of the consumer budgets. Moreover, the industry has a very large market to cater to, allowing them to charge higher prices. The existence of a large market is therefore crucial to the success of the industry.

In 2010, the World Bank's Global Consumption Database reported that the overall household expenditure on FMCG goods increased to US$240 billion in 39

African countries. Nigeria's household expenditure was equal to US$41.7 billion, followed by Egypt, which accumulated to US$27.6 billion, South Africa with US$23 billion, Morocco at (US$20.1 billion) and Ethiopia (US$19.2 billion). Similarly, other African countries with relatively improved FMCG markets identified include Ghana, Kenya, Tanzania, DRC and Cote d'Ivoire.

In Nigeria, there is a rising transformation in the FMCG industry as international supermarket brands flow into the country. According to the World Bank report in 2013, the Nigeria's household consumption expenditure was the highest in Africa (US$377 billion), meaning that the industry's performance and market are not constrained by consumers' income dynamics, as it consistently offers either necessities or accessible luxury goods in the country.

Moreover, given the recent occurrence of COVID-19 and decline in global oil prices worldwide, the growth in Nigeria's household expenditure has faced significant pressure, as the industry depends on the government policy tools to cushion the effects of the economic shocks. Similarly, the FMCG industry has been adversely impacted by the country's persistent rising inflation and interest rates, weaker currencies and tight fiscal policies. However, the trading up along the value chains has sustained the profit margins of the industry over the years, as many of their products continue to perform well in the Nigerian market. The Nigerian FMCG market is highly competitive and a larger percentage of the firms operating in the industry rely largely on downstream domestic industries (agriculture, manufacturing and agro-allied industries) to meaningfully deliver quality products to the customers. In Nigeria, the key players in the FMCG industry include the Nestle Nigerian Plc., Nigerian Breweries, Coca Cola, Vital Products, Friesland Campina Plc., PZ Cussons, Chi Limited, Unilever Nigeria, Dangote Group and others, while their products range from foods, drinks, detergents, beauty products and more.

Presently, there are a large number of FMCG firms in Nigeria, due to the new brands that are being introduced to the market daily, from which consumers can choose relatively. The presence of different products makes the industry more competitive and intense, as firms strive to produce quality products in the market place that would put them on the top of the list. This has also given the firms the edge to expand beyond the country to several parts of the Africa, as their products have gained wider popularity.

Generally speaking, the FMCG industry is classified into two major types depending on the number of products offered to the market place. In Nigeria, mono-brand firms offer only one type of product, like the Coca-Cola, while the multi-product firms offer more than one type of product, like Nestlé, Unilever and others.

The Role of Infrastructure Development in the Adoption and Diffusion of Internet-Based Technological Innovations in Nigeria

Physical infrastructure involves energy supply, transport systems (roads, railways, airways, waterways etc.) and telecommunications systems, among others. Physical infrastructure tends to function poorly in many parts of Nigeria, thereby leading in several instances to investment in alternative facilities or services such as electrical power generating sets, solar energy and power inverters among others. However, this tends to raise the production cost as well as the prices of goods and services, thereby discouraging demand. Also, a rise in the quality of infrastructure could greatly enhance productivity, but where such infrastructure is adverse or cost escalating, either the quality worsens or prices of utilities rise. Thus, technological infrastructure in this study is proxied by ICT infrastructure, which may be broadly divided into major components: telecommunications, connectivity infrastructure, computing and capabilities.

The persuasiveness of Internet-based technological innovations deployment is a function of acceptable costs to industrial firms and ease of usage, which depend critically on organizational technological capabilities and knowledge support systems in the industrial clusters. Similarly, the structure of the Internet service providers (ISPs) market is also a crucial determinant of Internet accessibility. The presence of ISPs and the institutional knowledge support systems in the industrial clusters they operate in is very critical to market competitiveness, and as such, entails a cost to end users.

Consequently, the diffusion of Internet-based technological innovations may be slow where Government policies create barriers to ISP free entry, or where operational limitations lead to persistent disparity among industrial firms, or where security concerns lead to hostility that prevents adequate competition. For instance, access at the individual level is attained using modems at the early stages of development, while more sophisticated infrastructure in this era of the 4IR would be required in later stages of development.

National, state and local telecommunications infrastructure (hardware) include server, local loop telecommunication lines, telecommunication mast, Wi-Fi routers, wireless LAN (WLAN), access points, switches, hubs, repeaters, gateways, bridges, modems, routers and edge platforms among others, while the software infrastructure entails monitoring and management tools, operating systems, apps, Domain Name System (DNS), Firewalls, Storage Area Networks (SAN), Load balancers and more. Firms in high-bandwidth telecommunications environments are most likely to have access to lower cost connections, but this might not be true in Nigerian industrial clusters due to poor and management of technological infrastructure over the decades. Most industrial firms face capacity constraints, due to low and thin bandwidth and erratic power supplies. Hence, the quality and capacity of the physical and technological infrastructure is vital to a firm's optimal performance.

However, the role of economic wealth and level of technological development becomes more important because the government and the industrial firms need significant level of investment capital for a reliable and efficient communications network, enabling them to experience and enjoy faster diffusion of the Internet-based technological innovations. National and global communications networks and connectivity are not possible without adequate economic wealth, especially in developing countries like Nigeria.

More importantly, the taxation level affects several elements of the connectivity chain, including investment decisions in Nigeria, and these variabilities are majorly caused by the breadth, complexity, uncertainty, unpredictability and varying levels of different taxes and contributions on the digital infrastructure and development, which often restrict infrastructure deployment and reduce affordability for consumers. As such, the use of complex and diverse taxes elements on Internet services needs to be balanced and focused on digital inclusion policy by setting appropriate policy priorities.

Therefore, the right investment in Internet network infrastructure can bring down consumer prices and improve affordability. For instance, before the commissioning of the SEACOM undersea cable in 2009, sub-Saharan Africa depended on the 340 Gbps SAT-3 cable systems for international connectivity and bandwidth, but between 2009 and 2012, the significant investment in seven new submarine cables improved the total capacity of the African continent to 25.7 Tbps. The number of submarine cable operating and connecting African continent has increased to 20, and there are plans underway to double the existing submarine cable capacity. Notably, the new capacity has transformed and changed the focus of the digital agenda in Eastern part of Africa, and Kenya, in particular with much higher service quality and reduced prices. Kenya has experienced short-term pricing terms as 1 megabyte of capacity dropped from USD 3000 per month to a range of USD 400 to USD 600 per month, while in longer term, pricing terms were as low as USD 150 to USD 200 per megabyte a month, and this pricing reduction led to considerable improvement in bandwidth usage rising from 2 Gbps in 2008 to more than 50 Gbps by the end of 2011.

Materials and Methods

The data for this study was obtained from a survey conducted in 2022. The survey was guided by the Organization for Economic Cooperation and Development (OECD) Third edition of the Oslo Manual: 'The Measurement of Scientific and Technological Activities'. The study employed a structured questionnaire to collect data from randomly selected fast-moving consumer goods (FMCG) industrial cluster firms. Sampled firms were drawn from the Nigerian business directory published by Manufacturer Association of Nigeria (MAN). Only industrial firms belonging to the International Standard Industrial Classification (ISIC) Rev.4 code 10–33 were selected for the study.

The study covered three states (Lagos, Ogun and Oyo) in South western Nigeria. The choice of these locations was informed by the predominance of industrial clusters in the states and their closeness to communications networks and infrastructural facilities. In addition, there is availability of learning, knowledge support system and the presence of physical infrastructure development in the clusters. Also, a multi-stage sampling method, which includes purposive, random and convenience sampling techniques was used in the selection of the industrial clusters' firms.

The first stage involves the purposive selection of the South-West from the six geo-political zones in Nigeria. The second stage entails selection of three states (Lagos, Ogun and Oyo) in Southwestern Nigeria, while the third stage entails the selection of the industrial clusters in these states. The fourth stage involves a random selection of fast-moving consumer goods (FMCG) firms in these clusters. Out of the 384 questionnaire distributed in the survey, 343 (89.3% response rate) were completed, retrieved and found useful for analysis. The response rates from selected industrial firms were 91.5% in Lagos, 89.5% in Ogun and 75.7% in Oyo state.

In order to ascertain the nature of the sampled industrial firms, they were asked to indicate whether they belong to either the public or private limited. The percentage of annual investment in Internet-based infrastructure in the last five years of operation was derived from the annual total IT expenditure of firms. The industrial clusters' firms were asked to ascertain whether they had engaged in any activities related to infrastructure development in their last five years of operations. The study adopted descriptive statistics, which included frequencies and percentages.

Results and Discussion

Approximately 85% of the FMCG firms sampled were private limited firms (Table 11.1). This is critical, as it could be an enabler and determinant for technological transformation and innovation, and has significant consequences on the financial supports and strength for new technology acquisition and infrastructure development. Notably, most of the firms operating in the Nigerian FMCG industrial clusters are private limited firms, indicating they are not listed in the stock exchange markets and cannot raise funds from the general public, except from the owners, friends and families. Thus, this could hinder the industrial firms from having adequate access to financial resources needed to thrive and compete. Also, most of the industrial clusters' firms (68.5%) spent less than 40% of their annual expenditure on Internet-based infrastructure development. This is an unacceptable level of investment as it could deter the industrial firms from performing optimally as expected.

Table 11.2 reveals the varying extent of infrastructure development among industrial cluster firms' extent of the infrastructure development in the implementation of Internet-based technological innovations and their capacities for technological absorption. Since the adoption and diffusion of any technology is influenced not only by industry-specific factors, but also by the level of a firm's technological infrastructure development. The Table shows that the most common infrastructure

Table 11.1 Nature and level of expenditure on Internet-based infrastructure among the Nigerian FMCG industrial clusters' firms

Respondents' characteristics	Frequency	Percentage (%)
Nature of firm		
Public Limited	52	15.2
Private Limited	291	84.8
Total	**343**	**100.0**
Percentage of IT annual expenditure on the Internet-based infrastructure		
Less than 20%	137	39.9
21% - 40%	98	28.6
41% - 65%	60	17.5
66% - 85%	45	13.1
86% and above	3	0.9
Total	**343**	**100.0**

development activities among the industrial clusters' firms are the acquisition of networking equipment (47.8%) and technological infrastructure (36.4%). However, the leasing or rental of equipment, as well as external linkages, are the least common forms of infrastructure development activities by these firms, with an average of 19% and 24% respectively.

This generally shows that the level of innovations for infrastructure development by the industrial firms to drive their productivity is relatively low. It becomes necessary for Nigerian industrial clusters' firms, in this reigning era of 4IR, to intensify

Table 11.2 Extent of the infrastructure development among the Nigerian FMCG industrial clusters' firms

Activities	Content	% of firms
Acquisition of technological infrastructure	Infrastructure or facilities readily available to firm for proper deployment of Internet-based services and its use to devise new and improved products and processes	36.4
Acquisition of networking	Acquisition of advanced networking equipment to produce new or significantly improved products, services and processes	47.8
Acquisition of software	Acquisition of software to produce new or significantly improved products, services and processes	31.7
Acquisition of technical	Acquisition of technical knowledge and skills knowledge and skills such as know-how to implement new or significantly improved products and services	27.7
External linkage	Internet infrastructure owned by your firm and managed by other firms (including other enterprises within your cluster) or by public or private organizations	24.4
Lease or rental of equipment	Lease or rental of Internet equipment or facilities to produce new or significantly improved products, services and processes	19.4

their innovative activities that would attract the attention of shareholders and policymakers to invest more in infrastructure development, so as to create a competitive advantageous environment for further improved performance and profitability. This implies that innovating firms in the industrial clusters should allocate their resources more efficiently and also leverage the learning and knowledge support system available in the clusters for effective infrastructure development. The industrial firms are encouraged to spend more on research and development (R&D) related to the improvement of their products and services, as well as enhancing the level of production processes to improve productivity. Any industrial firm that refuses to engage in R&D and innovation activities may find it very difficult to compete locally and globally with its rivals in the industry. More importantly, the Nigerian government can intensify efforts by providing tax rebate and other forms of incentives for the industrial firms that engage in technological innovations. Similarly, the right investment in Internet network infrastructure by the Nigerian government can bring down consumer prices and improve affordability.

Conclusion and Policy Implications

The gaps in innovations, technology and skills gaps among developed, developing and least developed economies have persistently and extensively widened because of huge asymmetry in access to new knowledge, capabilities, information, new technologies and modern infrastructure as well as technological learning and innovations channels. This study attempts to examine the extent of the infrastructure development in the implementation of Internet-based technological innovations in the Nigerian industrial clusters. The study suggests the need to strengthen and support the provision of infrastructure and innovation in the Nigerian industrial clusters, as this would increase firms' performance, promote infrastructural development and generate more employment opportunities. Technological progress in developing countries should place a high premium on local innovation infrastructure development and capabilities through the creation of new innovative activities and technologies, creating new opportunities for social and economic development, thus realizing higher levels of organizational, economic and social returns.

However, reflecting on the current state of infrastructure development in the Nigerian industrial sector, along with 'Industrialization-Infrastructure Nexus' in emerging economies, coupled with challenges and opportunities, leveraging innovations for infrastructure development, industrialization and promoting development cooperation through international partnerships are critical for sustainable industrialization. The industrial sector requires a critical nexus between technological innovations learning and Industry 4.0 to spur infrastructure development and boost manufacturing.

Critical issues such as poor infrastructure, lack of major services and support structures such as banking and financial services, productivity centres and training programmes, weak backward, horizontal and forward linkages and lack of timely information on local and foreign markets tend to reinforce the low growth dynamic observed in the clusters. Thus, for Nigerian industrial firms not to be at a serious risk of being left behind there is an urgent need for technological infrastructure revolution that can transform the sector. Attaining vitality in the long-term for the cluster depends on its ability to attract companies, venture capital, skills, infrastructure, technologies and other resources from all over the world, in what is termed the 'Greta Garbo-effect'. However, this would create much of the technological infrastructure needed to accelerate the development of the network, but the deploying and usage of infrastructure remains an economic problem due to the high rollout costs and low returns. Therefore, the right investment in Internet network infrastructure can bring down consumer prices and improve affordability. The digital divide continues around the world in variety of different forms, particularly gaps in universal accessibility, with high-speed Internet connectivity infrastructure even in countries and the COVID-19 pandemic has brought further attention to the disparities in digital infrastructure, digital skills and safety and security as the world has been forced to live, work and learn remotely. The effects of COVID-19 may also have widened the digital divide within individual countries as it has particularly entrenched differences in network infrastructure, readiness, safety and access to connectivity in both developed and developing economies. Similarly, the provision of technological infrastructure is crucial to support industrial firms' performance, which would improve their ability and potential to compete in international markets through the acquisition and application of the new technologies. There is a need to intensify greater infrastructure supports so as to fully reap and benefit from new technologies and develop capabilities to promote economic development.

References

Akinwale, Y. O., Adepoju, A. O., & Olomu, M. O. (2017). The impact of technological innovation on SME's profitability in Nigeria. *International Journal of Research, Innovation and Commercialisation, 1*(1), 74–92.

Amobi, I. (2016). Unleashing industrial clusters for growth and prosperity in South Eastern Nigeria. In *Proceedings from the Enugu Forum Seminar, October*.

Azam, M. (2015). Diffusion of ICT and SME Performance. In *E-Services Adoption: Processes by Firms in Developing Nations. Advances in Business Marketing and Purchasing, 23*(A), 7–290. Accessed December 14, 2022 https://doi.org/10.1108/S1069-096420150000023005

Bank of industry. (2018). Industrial clusters and economic development in Nigeria. Working Paper Series, No. 4, 27th November. Accessed October 28, 2022 https://www.boi.ng/wp-content/uploads/2019/02/BOI-Working-Paper-Series-No4-Industry-Clusters-and-Economic-Development.pdf

Chete, L. N., Adeoti, J. O., Adeyinka, F. M., & Ogundele, O. (2013). Industrial development and growth in Nigeria: Lessons and challenges. Learning to compete, Africa growth initiative at Brookings (AGI), the African development bank, (AfDB), and the United Nations University World Institute for development economics research (UNU-WIDER) on industrial development in Africa, Working paper No. 8.

Ekesiobi, C., Ude, D., & Nwokolo, C. (2018). Industrial clusters and industrialisation in Nigeria: A Micro-assessment of the Nnewi automotive component industrial cluster, Anambra State. *The Nigerian Journal of Economic and Social Studies, 60*, 131–162.

Government of Republic of Korea. (2020). Infographic of 2020 annual report on national informatization. Accessed December 15, 2022 https://www.msit.go.kr/SYNAP/skin/doc.html?fn=5b528f8a2ce056dc686fef46eb8671ad&rs=/SYNAP/sn3hcv/result/

GSMA. (2020). The state of mobile internet connectivity. Accessed on 02/04/2021. Available at: https://www.gsma.com/r/wp-content/uploads/2020/09/GSMA-State-of-Mobile-Internet-Connectivity-Report-2020.pdf

Hong, E.-K., Ryu, J. M., & Lee, E. J. H. (2021). Entering the 5G era: Lessons from Korea. World Bank Group Korea Office Innovation and Technology Note Series No. 5. Accessed December 14, 2022 https://www.worldbank.org/en/country/korea

Humphrey, J., & Schmitz, H. (2000). Governance and upgrading: Linking industrial cluster and global value chain research. Brighton, IDS Working paper 120.

Kaplinsky, R., & Readman, J. (2001). Integrating SMEs in global value chains, towards partnership for development. UNIDO, Vienna.

Kongmanila, X., & Takahashib, Y. (2009). Inter-firm cooperation and firm performance: An empirical study of the Lao Garment industrial cluster. *International Journal of Business and Management, 4*(5), 3–17.

Lee, H.-J., Jeong, C. H., Xu, W., & Fairhurst, A. (2014). The influence of consumer traits and demographics on intention to use retail self-service checkouts. *Marketing Intelligence and Planning, 28*, 46–58.

De Marez, L., Evens, T., & Stragier, J. (2011). Diffusion theory versus today's ICT environment. *Observatorio Journal,5*(3), 175–202. Accessed January 14, 2021. http://obs.obercom.pt

Naudé, W., Paula, N. (2015). Industrialisation, innovation inclusion, UNIDO inclusive and sustainable development. Working paper series, research, statistics and industrial policy branch WP 15, October 30.

Olomu, M. O., Binuyo, G. O., & Oyebisi, T. O. (2023). The adoption and impact of Internet-based technological innovations on the performance of the industrial cluster firms. *Journal of Economy and Technology, 1*, 164–178, ISSN 2949–9488. Available at: https://doi.org/10.1016/j.ject.2023.11.004

Osarenkhoe, A., & Fjellström, D. (2017). Clusters' vital role in promoting international competitive advantage–Towards an explanatory model of regional growth. *Journal of Regional Research, 39*, 175–194.

Oyelaran-Oyeyinka, B. (2001). Networks and linkages in African manufacturing cluster: A Nigerian case study. The United Nations University, Discussion paper series No. 5, Helsinki.

Pandit, N. R., Cook, G. A. S., Wan, F., Beaverstock, J. V., & Ghauri, P. N. (2018). The economies and diseconomies of industrial clustering: multinational enterprises versus uninational enterprises. *Management International Review, 58*, 935–967.

Roblek, V., Meško, M., Bach, M. P., Thorpe, O., & Šprajc, P. (2020). The interaction between internet, sustainable development, and emergence of society 5.0. *Data, 5*, 80. https://doi.org/10.3390/data5030080

Somuyiwa, A. O., & Oyesiku, O. O. (2010). Analysis of performance of collaborative sharing cost in outbound logistics. *Current Research of Social Sciences-Maxwell Scientific Organisation, 2*(3), 181–186.

Ugoani, J. N. N., & Ibeenwo, G. I. (2015). Entrepreneurship development and employment generation in Nigeria: a study of the national directorate of employment. *Independent Journal of Management and Production, 6*(3), 687–710.

United Nations industrial development organisation (UNIDO). (2022). Advancing economic competitiveness and supporting small and medium industry clusters. Accessed February 25, 2022 https://www.unido.org/our-focus/advancing-economic-competitiveness/supporting-small-and-medium-industry-clusters

Uzor, O. O. (2014). Small and medium scale enterprises cluster development in South-Eastern Region of Nigeria. Institute for world economics and international management, Berichte aus dem Weltwirtschaftlichen Colloquium der Universität Bremen, Nr. 86, ISSN 0948–3829.

Venkatesh, V., Thong, J. Y., & Xu, X. (2012). Consumer acceptance and use of information technology: Extending the unified theory of acceptance and use of technology. *MIS Quarterly, 36*. 157–178.

World economic forum. (2016). Global information technology report: Innovating in the digital economy. Geneva. Accessed March 20, 2021 http://www3.weforum.org/docs/GITR2016/WEF_GITR_Full_Report.pdf

Zhang, X. (2018). Frugal innovation and the digital divide: Developing an extended model of the diffusion of innovations. *International Journal of Innovation Studies, 2*, 53–64. Accessed August 12, 2021 https://doi.org/10.1016/j.ijis.2018.06.001

Chapter 12
World's Energy Transition Dilemma and Its Implications for Industrialisation—Observations from South Africa

Thokozani Simelane, Christian Green, and Thulisile Mphambukeli

Abstract The global goal to transitioning to renewable energy has created a dilemma for the international community. This, however, has the potential to slow the industrial capabilities of developing countries. From South Africa's experiences, most developing countries lack the necessary resources, such as the large sums of money required to build the infrastructure, needed for a comprehensive shift to renewable energy. Furthermore, the world's industrial infrastructure is still geared toward using fossil fuels as its primary source of energy. Given this reality, it is critical to underline that, in order to sustain human development, the energy transition must have a minimal impact on countries' industrial capacities and development, particularly in developing countries. To mitigate the potential negative consequences of the energy transition, developing countries should not rush into the renewable energy bandwagon without first thoroughly examining and understanding the ramifications, particularly for their industrialisation capability and infrastructure development. South Africa's experience with renewable energy serves as an example, particularly for the developing countries. Through the development and implementation of renewable energy policies, as well as investments in renewable energy technologies, the country has demonstrated that it is possible to transition away from fossil fuels toward a sustainable renewable energy future. This paper offers recommendations for energy transition options that developing countries should examine. It outlines

T. Simelane (✉)
Faculty of Engineering and Built Environment, University of Johannesburg and Human Sciences Research Council, Urban and Regional Planning, Johannesburg, South Africa
e-mail: tsimelane@uj.ac.za

C. Green
Department of Electricity and Energy, Pretoria, South Africa
e-mail: christian.green@dee.gov.za

T. Mphambukeli
Faculty of Engineering and Built Environment, University of Johannesburg, Urban and Regional Planning, Johannesburg, South Africa
e-mail: tmphambukeli@uj.ac.za

X. Poshiwa et al. (eds.), *Technological Innovations for Sustainable Industrialization*,
https://doi.org/10.1007/978-981-96-7995-9_12

the history of South Africa's electrical supply, demand and production. It looks into how prior energy policies arose, assesses their benefits and drawbacks, and examines how they matured over time. The goal is to contribute to an understanding of the industrial effects of the transition to renewable energy sources, which is being considered globally.

Keywords Renewable energy · Energy transition · Solar · BRICS+ · South Africa

Introduction

Despite international calls to shift toward renewable energy sources (Issa et al., 2023; Tawalbeh et al., 2024) the world's industrial infrastructure is still designed for energy supplied primarily by fossil fuels (Hassan et al., 2020). This means that the world is facing a significant dilemma regarding energy transition (Wang & Li, 2022). Most developing countries are lagging behind in implementing renewable energy infrastructure capable of powering industrial processes and systems (Dada et al., 2023). They also lack adequate resources, such as the substantial financial capital required to construct the infrastructure essential for a comprehensive transition to renewable energy (Zhang et al., 2021).

Given this reality, it is essential to emphasise that, to sustain human development, the emerging energy transition must exert minimal impact on countries' industrial capacities and development, especially within developing countries (Bhattacharya et al., 2016). According to studies on the socioeconomic benefits of fossil fuel-based energy industries, the era of fossil fuel energy sources has had a profound impact on human development (Fang et al., 2020). The industrial contribution of fossil fuels to social and human development is clearly evidenced by the chemical and energy industries that have arisen from fossil fuel exploration and exploitation (Wolde-Rufael, 2009). They use the energy content of naturally occurring substances, such as liquid and gaseous hydrocarbons, across most energy-generation value chains. They also utilise the molecular composition of these resources, which comprises a complex array of organic compounds specifically developed as feedstock for numerous chemical processes and industrial products within the chemical industries. All these combined constitute a complex industrial ecosystem that is entirely based on the exploitation of fossil fuels for energy. As a result, industrial infrastructure and processes that either supply raw materials or utilise by-products of refined fossil fuels have been developed over the years. These systems cannot be easily converted or substituted overnight.

With an accelerated migration to renewable energy sources, which a large portion of the world advocates for, developing countries must recognise that, as a transition to renewable energy becomes imminent, careful consideration must be given to the current industrial implications, economic and social benefits of fossil fuels (Setiawan & Anwar, 2022). Therefore, the transition to renewable energy should not be unduly disruptive. South Africa and other developing nations such as China,

India, Brazil, the United Arab Emirates, Pakistan, and others may serve as intriguing case studies to observe and demonstrate the potential, both expected and unexpected, consequences of switching to renewable energy sources without adequately considering industrial costs and other ramifications into account (Okorie et al., 2018).

South Africa has a long history of relying heavily on coal-fired power plants to provide electricity. The country's large coal deposits and the relatively low price of coal-based electricity have fuelled South Africa's dependence on coal (Kaggwa et al., 2011), which is currently being reassessed.

For years, the nation's expanding energy needs have been successfully met through coal generated electricity because it was both readily available and extensively mined. In its dependency on coal, South Africa's energy sector has developed a complicated history shaped by a variety of factors, such as economic growth, technological advancements and environmental concerns. As such, South Africa has a track record of producing both conventional and alternative energy (as exemplified by SASOL), and it has long been seen as a significant player in the energy industry (Abdel-Rahman & Simelane, 2011).

Due to climate change and other environmental concerns (Usman & Balsalobre-Lorente, 2022), South Africa's historical reliance on coal is no longer considered sustainable. There is a widespread perception that the time has come for South Africa to switch from its heavy reliance on coal-based electricity and transition to renewable energy sources (Amigun et al., 2011).

This situation has created an unavoidable energy policy dilemma in South Africa's energy sector, as the government strives to find a balance between three policy objectives i.e. security and accessibility, short- and medium-term affordability, and environmental sustainability. As a result of this dilemma, it is currently widely contested whether South Africa should completely transition to renewable energy sources to meet the majority of its energy needs. However, South Africa maintains that, as a developing country, there are other energy options worth considering.

To safeguard its industrial capacity, the country advocates for nuclear power as part of its energy mix, despite the fact that renewable energy sources such as solar, wind and hydropower (especially from the ocean) are considered viable alternatives. The country believes that its approach is consistent with the strategies adopted by the developing countries, such as Indonesia, China and India, as well as developed nations like France (Hermawan and Sudjadji, 2022).

According to South Africa, its energy mix provides the country with the ability to preserve its industrial production capacity, grow economically, maintain high industrial output to sustain the economy and ultimately achieve the level of human development it aspires to. While South Africa's energy dilemma continues to be debated, it must, as a developing country, pursue a well thought out energy path. This chapter presents perspectives on potential energy pathways that developing countries like South Africa can consider.

The aim is to contribute to an understanding of the industrial implications of global shift toward renewable energy sources, a transition that is being explored worldwide. This chapter places particular emphasis on South Africa's electricity sector. It recounts the history of electricity supply, demand and production in South

Africa. The paper examines the evolution of past policies, evaluates their strengths and weaknesses and analyses how they have developed over time.

Finally, the paper argues for what BRICS+ bloc, as a group of developing countries, can make a meaningful contribution to addressing the world's energy transition dilemma. It concludes by suggesting a viable course of action that should be considered by developing countries.

Tracing South Africa's Energy Dilemma

The turning point in South Africa's energy sector occurred on April 27, 1994. This event marked a fundamental shift in the country's energy and economic trajectory. The newly elected government reviewed existing public policies and established new policy imperatives aimed at transforming South Africa into a nation where all citizens benefit from and contribute to the country's economic progress.

One of these priorities was the recognition of the importance of expanding electricity access to all citizens. It is worth noting that this commitment was made under the assumption that Eskom the sole entity responsible for electricity generation and distribution would be able to continue supplying power and meeting the demand of all households connected to the grid. This is despite the age and legacy of coal-powered power stations, which political authorities should have taken into consideration.

Eskom was founded in the late nineteenth century, when Cape Town's first power plant was erected, marking the inception of South Africa's energy production history. This represented the first step in electrifying and modernising South Africa's economy. Since then, Eskom has been the sole entity responsible for producing all of South Africa's electricity. For South Africa, coal has remained the primary source of electricity. It was easily accessible and affordable, making it the preferred choice for powering industries and meeting the country's energy demands. However, as the negative effects of coal on the environment and human health have become more apparent and international pressure has mounted, South Africa has found itself in an energy dilemma and has started to explore new sources of energy.

This includes thoroughly assessing an energy mix that incorporates renewable sources of energy. Additionally, the use of nuclear energy is under careful consideration. To facilitate this, national programmes have been established involving the collaboration among multiple institutions to engage in long-term research on the complex subject of energy transition in South Africa.

Collaboration has been forged between academic research institutions like the Council for Scientific and Industrial Research (CSIR), MINTEK, SASOL (a South Africa based international petrochemical company), Universities, the Chamber of Mines, and several other stakeholders. The Government White Paper on Energy Policy, published in 1998, laid out a plan for reorienting and liberalising the energy delivery sector. However, political and executive leadership at the national, provincial and local levels along with the industry leaders and various affected stakeholders soon became clearly divided on the issue. Incumbents in the electricity supply industry, such as Eskom and municipalities in particular, showed little willingness to implement the recommendations of the 1998 White Paper on Energy Policy.

As a result of this impasse, South Africa is facing an energy crisis due to its inability to align energy policies, execute recommendations, invest in new power infrastructure and properly maintain existing facilities. Meanwhile, energy demand has increased significantly, due to increasing access to electricity and ongoing industrialisation.

Milestone in South Africa's Energy Transition

A significant milestone for South Africa's energy transition is reflected in its policy commitments to shift toward renewable energy. One such milestone is the establishment of a target to substantially increase the share of renewables in energy mix by 2030. This target underscores South Africa's commitment to expanding its renewable energy capacity and reducing its reliance on fossil fuels, particularly coal. In pursuing this path, the country has faced a challenging and uncertain trajectory of integrating renewable energy into its national energy mix.

This has not only transformed the country's energy landscape, but has also raised concerns about its industrial development and broader economic growth. Overall, renewable energy in South Africa is essential for addressing the country's electricity shortages. It not only diversifies the energy mix by reducing reliance on a single fuel source, but also has the potential to reduce greenhouse gas emissions and air pollution. This diversification will enhance energy security while reducing the risks associated with volatile fossil fuel prices. Another important milestone is the integration of renewable energy resources into the national power grid. This integration has been enabled by the enactment of energy policies implemented by the South African government, which promote the use of clean energy.

The Renewable Energy Independent Power Producer Procurement Programme (REIPPPP) is one such initiative that has been successful in attracting private investment and promoting the deployment of renewable energy projects. Through this programme, South Africa has significantly increased its renewable energy capacity. As of 2021, the country successfully met its targets for generating renewable energy and diversifying its energy sources. This is commendable progress, but there is still much work to be done regarding the investment needed to develop the required infrastructure and to improve the reliability.

South Africa's renewable energy policies have not only promoted the adoption of renewable energy sources but have also spurred the unbundling and deregulation of the energy sector, creating a more favourable environment for the growth of renewable energy initiatives. As such, renewable energy has helped to reduce the negative economic, social and environmental consequences of energy production and consumption. It has further created opportunities for skill development and investment in energy sector.

As currently widely recognised, adopting renewable energy has the potential to improve electricity access, an ambition that the ruling party has maintained since taking office. By diversifying its energy sources and reducing reliance on imported fossil fuels, South Africa can ensure a more stable and secure energy supply for all of its residents. Furthermore, adopting renewable energy could foster scientific advancements and innovation through collaboration with other emerging countries.

Technological Advancements in South Africa's Energy Sector

The shift to renewable energy sources can be greatly accelerated by technological developments in South Africa's energy sector. By intensifying investment in research and development, South Africa can create new, more efficient, accessible and affordable renewable energy solutions for all. These developments may enhance the effectiveness of renewable energy sources, providing a competitive alternative to the current fossil fuel-based energy industry.

For a time, government institutions such as Institutions of Higher Learning, SANEDI (South African National Energy Development Institute), the CSIR (Council for Scientific and Industrial Research) and others have played the role in initiating projects of strategic significance for South Africa's future energy mix. Below are examples of some programmes and research projects conducted to date.

Nuclear

The Minister of Mineral Resources and Energy and the nuclear industry in South Africa appear to be placing their hopes in the deployment of small modular nuclear reactors, despite these designs not yet being finalised, built, licensed, piloted or proven safe and economically viable for commercial power generation.

This follows a heated debate on whether South Africa should pursue the integration of nuclear energy into its energy strategy. The Integrated Resource Plan (IRP) made provision to start planning for a nuclear build programme of up to 2500 MW at a pace and scale that the country can afford. However, it should not be inferred from this IRP provision that a new nuclear build will actually occur in South Africa in the near future.

Gas

In May 2016, the government released a draft Gas Utilisation Master Plan (GUMP) for public discussion. However, the final version of GUMP was never made available to the general public. As a result, preparations for importing liquefied natural gas, compressed natural gas, liquefied petroleum gas, and shale gas remain disorganised and uncoordinated. The current Minister of Mineral Resources and Energy frequently emphasises that gas is a game-changer in South Africa and has shown interest in the recent significant discoveries of gas condensate made by total approximately about 175 kms off the southern coast of South Africa, as well as the substantial developments of gas fields in Mozambique.

Renewable

Wind and solar PV make up the vast majority of new construction under the IRP 2019 for the years through 2030, supported by flexible generation options as gas-to-power and battery energy storage. This reflects the fact that this is, in fact, the preferred alternative because it provides consistent, dispatchable energy at the lowest cost, with the least water usage and emissions, while also generating the most jobs and supporting South Africa's climate change obligations.

There is also a growing realisation, including by Eskom itself, that the utility-scale renewable energy build programme in IRP 2019 has to be significantly expanded even further in to meet the falling output from Eskom' aged coal-fired power stations, due to decreasing availability and pressure from lenders for an accelerated decommissioning and decarbonisation schedule. However, the uptake of renewable energy detailed in IRP 2019 is constrained by policy interventions in the form of artificial and arbitrary limits on the annual deployment of wind and solar PV power plants of 1600 megawatts per annum for wind and 1000 megawatts per annum for solar PV.

Hydrogen

At this stage, there is no formal, published South African government policy or strategy yet on the role and production of green hydrogen. However, the Hydrogen South Africa (HySA) strategy and Hydrogen Society Roadmap are the works in progress under the auspices of the Department of Science and Innovation, the Council for Scientific and Industrial Research, Mintek and several universities. There is growing recognition that South Africa's production of green hydrogen and green ammonia provides major competitive advantages for local decarbonisation of transport and industry, as well as for export. These include plentiful wind and solar natural resources, inexpensive, available land and over 50 years of commercial experience producing synthetic fuels using the Fischer–Tropsch method, which SASOL uses to manufacture fuel from coal.

Challenges in South Africa's Energy Transition Path

Despite the numerous benefits of renewable energy, South Africa still faces several challenges in its transition to clean electricity production. One of the major challenges is the high upfront of implementing renewable energy infrastructure. Investing in renewable energy projects requires substantial upfront capital, which can be a barrier for developing countries like South Africa. Another challenge is the integration of renewable energy into the existing grid infrastructure. Updating and upgrading the grid to accommodate a larger share of renewable energy can be a complex and costly process. Overall, the evolution of energy and electricity production in South Africa is critical for its future economic, social and environmental prosperity. Investment in renewable energy and energy efficiency is essential for South Africa to reduce the negative economic, social and environmental impacts of energy production and consumption (Lombard & Ferreira, 2015).

This transition will not only address the imminent energy crisis which was highlighted over a decade ago, but also contribute to global efforts in mitigating climate change and meeting sustainable development goals. Investment in renewable energy and energy efficiency is crucial to reduce the negative economic, social and environmental impacts of energy production and consumption in South Africa (Lombard & Ferreira, 2015). South Africa has started to make significant progress in embracing renewable energy through the implementation of a renewable energy policy (Nape et al., 2019).

South Africa must maintain its dedication to creating a robust renewable energy policy in order to overcome these obstacles and advance the transition to renewable energy. This entails offering financial assistance and incentives for renewable energy projects, as well as establishing a supportive regulatory environment that promotes investment in the industry.

In order to spur innovation and find affordable solutions, South Africa should promote research and development in renewable energy technologies. Through this, South Africa can create new, more efficient renewable energy technologies that are both cost effective and accessible to everyone. It can be challenging and costly to make the switch to using more renewable energy. In order to move toward a more sustainable energy future, South Africa must maintain a balance between the energy supply needed to support industrial activity and the expectations from other countries for a sudden switch to renewable energy sources. This indicates that an energy paradigm shift is essential in South Africa, since the country's reliance on coal-based energy generation is under attack from many directions and can no longer be sustained.

Future Prospects of South Africa's Energy Transition

The future prospects of energy and electricity production in South Africa are promising. With the country's commitment to renewable energy and the implementation of policies such as the Renewable Energy Independent Power Producer Procurement Programme (REIPPPP), South Africa is well on its way to achieving its renewable energy targets (Nape et al., 2019).

These targets include generating 20,000 MW of renewable energy by 2030. This commitment to renewable energy will not only help South Africa meet its growing energy demands, but also position the country as a leader in sustainable and clean energy production. Additionally, South Africa's collaboration with other BRICS countries to develop its renewable energy sector further highlights its commitment to a greener, more sustainable future (Ndlovu & Inglesi-Lotz, 2019).

The integration of renewable energy into South Africa's energy mix not only benefits the environment, but also presents substantial economic opportunities. The renewable energy sector has the potential to create thousands of jobs, stimulate economic growth and attract both domestic and foreign investment. Investing in renewable energy and energy efficiency is not just a choice, but a necessity for South Africa's future. By embracing renewable energy, South Africa can reduce its reliance on fossil fuels, diversify its energy sources and create a more sustainable and resilient energy system that can withstand future energy shocks or supply disruptions. Furthermore, transitioning to renewable energy will also improve the country's air quality and public health. Investing in renewable energy and energy efficiency is crucial for reducing the negative economic, social and environmental impacts of energy production and consumption in South Africa (Lombard & Ferreira, 2015).

South Africa must prioritise investment in renewable energy and energy efficiency to reduce the negative economic, social and environmental impacts of its current energy production and consumption (Lombard & Ferreira, 2015). The coal mining industry itself produced the so-called National Coal Strategy for South Africa in 2018, to determine what needs to be done in the face of negative public opinion regarding the use of coal in power generation and industrial processes. However, there is a growing awareness within government, among the coal mining majors, and within the financial sector that coal mining in South Africa is a declining industry under threat. Despite being well behind the curve as one of the most carbon-intensive economies in the world, the realisation of the need for decarbonising of the economy is beginning to gain traction.

However, distracting talk of 'clean coal' technologies, underground coal gasification and carbon capture and storage is still heard emanating from the Department of Mineral Resources and Energy (DMRE) and other quarters, demonstrating a lingering commitment to coal as a source of the primary energy.

Role of Government in South Africa's Energy Transition

Between 1994 and 1999, South Africa produced more than 800 acts of parliament to drive a modernisation agenda. These acts included, among others, the Labour Relations Act (1996), the Public Finance Management Act (1999) and the Employment Equity Act (1998). These three pieces of legislation were aimed specifically at improving labour relations and institutional governance.

The Higher Education Act (1999) provides the framework for the reorganisation of higher education through a series of institutional mergers. In theory, the Immigration Act (2002) provides a mechanism to address scarce skills through foreign recruitment, but in practice, it erects administrative barriers that are likely to impede inward flows. In 1998, a White Paper on Energy policy was developed to initiate a process for the transition of South Africa's energy and electricity industry. However, at the time, the political economy was not supportive of the recommendations.

More than 20 years later, the work that was initiated with the 1998 White Paper on energy policy provided the platform for the current debate on a roadmap for Eskom and the Electricity Supply Industry. A selection of policy statements extracted from that White Paper, which are pertinent to the matters that are dominating the current discussion that informs the vision for Eskom and the ESI, include:

(i) In the long term, Eskom will have to be restructured into separate generation and transmission companies. (ii) The entry of multiple players into the generation market will be encouraged (The purpose of this policy is to develop a competitive power market). (iii) Government will legislate for transmission lines to provide for non-discriminatory open access to uncommitted capacity, transparency of tariffs, and disclosure of cost and pricing information to the National Electricity Regulator. (iv) Licensed privately owned distributors will be allowed to co-exist alongside other public and private distributors to distribute their own generated electricity, subject to approval by the National Electricity Regulator. Given these key policy statements, the White Paper emphasises;

(i) Giving customers the right to choose their electricity supplier; (ii) introducing competition into the industry, especially in the generation sector; (iii) permitting open, non-discriminatory access to the transmission system; and (iv) encouraging private sector participation in the industry. The key point the White Paper makes is that competition should be enhanced in all sectors, except transmission. Although the key policy statements of the White Paper were made in 1998, it still resonates with the current requirements, specifically with regard to the developments in the global energy sector.

Accordingly, the vision for South Africa's Electricity Supply Industry and that of Eskom, for the next 1–520 years, will also be informed by the fact that the energy sector globally is transitioning, driven primarily by technological developments and the need for lower carbon emissions.

The White Paper laid the foundations for the 2019 'Roadmap for Eskom in a Reformed Electricity Supply Industry'. It has its main focus on the restructuring of Eskom into three subsidiary businesses, generation, transmission and distribution in an effort to assist with adequate power supply. It is recognised that the electricity reform process will take place over several years to ensure that South Africa has an appropriate electricity supply system capable of responding to the demands of a rapidly evolving world. In the reform process of Eskom, the diversification of Eskom's primary energy sources towards renewable energy sources will be done in a 'Just' manner. This will entail alternative economic activities to be developed and implemented to economically sustain workers and communities dependent on power stations that will be decommissioned.

Contribution of BRICS+ in Assisting the World's Energy Transition

Brazil, Russia, India, China and South Africa, along with Egypt, Ethiopia, Iran, the United Arab Emirates and Saudi Arabia, comprise a diverse group of countries with indisputable global energy clout, both in terms of production and demand. They form an economic bloc whose energy production and demand have a significant impact on world growth (Simelane & Knop, 2022). As a developing-country bloc, BRICS+ must take the chance to lead the development direction of energy transition, making it easier for developing countries to strike a balance between continuing industrial development and gradually transitioning to renewable energy sources (Kahia et al., 2021). The worldwide involvement of the BRICS+ in energy, both generation and consumption, impacts the global energy market's trajectory. This has an impact on the world's energy security and industrialisation.

Energy is the fundamental component that drives economic development and growth; hence there is a strong link between economic growth, energy consumption and industrialisation. With their natural resource reserves, the BRICS+ countries contribute significantly to the worldwide supply of energy and its dynamics, which have a direct impact on global industrial development. The BRICS+ countries have the greatest influence on global energy supply and demand. This is the foundation for moving to development powered by renewable energy technology.

As an example, both Saudi Arabia and the United Arab Emirates are among the World's producers of hydrocarbons and hold records of having the largest oil reserves, while China is the largest producer of hydropower, solar electricity and wind power among the BRICS+ countries. In terms of renewable energy output, India is second only to China. Brazil is the world leader in biofuel production, which it combines with hydropower, wind and solar energy. South Africa, like the other emerging economies within its economic group, is contending with reorienting its economy towards renewable energy sources that might weaken its industrial capacity. The country has taken caution in the transition towards renewable energy sources by developing a series of frameworks and action plans. The country is slowly incorporating renewables into its energy mix. It has created a comprehensive integrated resource strategy, which seeks to balance industrial development and domestic energy demand.

Implications of Energy Transition to Industrialisation of Developing Nations

The transition to renewable energy must be gradual rather than abrupt in developing countries. Co-development of technology among developing countries is critical for improving collaboration (Simelane et al., 2023). In terms of the energy-development nexus, developing countries must rethink their energy positions in order to sustain their industrial development. This can be accomplished by unified policies, sharing of energy-generation technologies, a commitment to contribute to the global carbon-reduction programmes and industrial growth routes that meet the economic needs of developing countries.

On a global scale, BRICS+ has shifted power dynamics, making the bloc a key player in solving the World's energy dilemma. The BRICS+ countries are highly represented in all energy sectors that drive global industrial development. They have a high level of energy security, and as a result, a framework for collaboration in developing a model for energy transition exists.

BRICS+ countries have the expertise and technologies to exploit their energy resources to improve the energy security of the World. This can occur in all energy sources, including oil and gas, renewable energy and nuclear energy. Energy, whether used or generated, is a crucial economic stimulus for industrial development, and collaboration among developing nations is desirable. As a result, affordable, reliable and ideally sustainable energy production that will sustain industrial development among developing nations is critical for them to realise their development goals and aspirations.

Concluding Remarks

South Africa's journey towards a sustainable and renewable energy future serves as an inspiration for the rest of the world. By implementing renewable energy policies and investing in clean energy projects, South Africa is showing that it is possible to transition away from fossil fuels and embrace a green energy future (Nape et al., 2019). This transformation not only of benefit to the environment but also creates new economic opportunities, stimulates growth and improves public health. Investing in renewable energy and energy efficiency is not just a choice, but a necessity for South Africa's future. By prioritising renewable energy and embracing clean technology, South Africa is setting an example for other countries to follow. The evolution of energy and electricity production in South Africa has been modified by the recognition of the urgent need to expand access to electricity to all citizens of the country. By implementing initiatives like the Renewable Energy Independent Power Producer Procurement Programme and participating in international agreements like the United Nations Framework Convention on Climate Change's Clean Development

Mechanism, South Africa is not only aligning its stance on energy access, but also supporting the international agreements, programmes and developments.

The amalgamation of the Department of Mineral Resources with the Department of Energy in early 2020 to form the Department of Mineral Resources and Energy created the need for significant structural and people change-management, which slowed the work of the department in the energy sector, and also created some conflicts of interest between serving coal mining interests and the need to decarbonise the South African economy. Within the department, there is still continuing talk of clean-coal technologies, underground coal gasification, carbon capture and storage and futuristic nuclear power technologies that do not provide economically viable, flexible and proven solutions to electricity generation capacity shortfalls in the short but may benefit the country in a medium to long term.

There are indeed some positive and pragmatic statements and policy positions coming from political leaders and government. Some examples include: the restructuring and unbundling of Eskom, which is inevitable; the need for regulatory reform and the easing of regulatory red tape; the establishment of an independent transmission system and market operator; the opening of the power generation sector to municipal generators, distributed generation, self-generation and wheeling; and the need for decarbonisation of the economy and a just energy transition.

To maintain its industrial capacity, South Africa's government must provide clear, unambiguous and consistent policy positions and messaging to attract investment and encourage private sector participation in energy generation, storage and distribution. The platform provided by the need for developing countries to approach the transition to renewable energy and predict its effects on industrial capacity allows South Africa to align its programmes and contribute to the global energy transition dilemma by assisting developing countries in developing agile and effective energy transition policies and programmes.

At an international level, Pathak and Shah (2019) argue that energy security is the foundation for every nation's socioeconomic prosperity. The BRICS+ group of countries plays a key international role in both the generation and use of global energy resources. This has the ability to shape the path of the world's energy transition. The BRICS+ countries are thus a dynamic mix of growing economies whose reliance on energy provides chances, strengths and many choices for closer collaboration among developing countries (Russian Federation's Ministry of Energy, 2017).

However, as the world becomes more polarised and energy is used as a weapon in international conflict, it is vital to assess the impacts of transitioning to renewable energy sources through investments and mergers among developing countries. This could demand greater collaboration in innovation and technology transfer via established mechanisms such as the Kyoto Protocol, as not all developing countries have equal capacity to develop technologies (Simelane & Inyang, 2023). Despite this promise, there is growing concern that emerging countries should not simply jump on the renewable energy bandwagon without thoroughly assessing and understanding the implications, particularly for their industrialisation and capacity.

References

Amigun, B., Kaggwa, M., Musango, J., Mutanga, M., Simelane, T., & Williams, S. (2011). Africa's technology options for renewable energy production and distribution. In: T. Simelane, M. Abdel-Rahman (Eds.), *Energy transition in Africa*. AISA. Pretoria.

Bhattacharya, M., Paramati, S. R., Ozturk, I., & Bhattacharya, S. (2016). The effect of renewable energy consumption on economic growth: Evidence from top 30 countries. *Applied Energy, 162*, 733–714. https://doi.org/10.1016/j.openenergy.2015.10.104

Dada, M, Popoola, P, Alao, A, FolasayoOlalere, F, EvllyMtileni, E, NtanziLindokuhle, N, & Shamaine, M. (2023). Functional materials for solar thermophotovoltaic devices in energy conversion applications: a review. *Frontiers in Energy Research, 11*. https://doi.org/10.3389/fenrg.2023.1124288

Fang, G., Tian, L., & Yang, Z. (2020). The construction of a comprehensive multidimensional energy index. *Energy Economics*. 104875. https://doi.org/10.1016/j.eneco.2020.104875

Hassan, A. A., El Habrouk, M., & Deghedie, S. (2020). Renewable energy for robots and robots for renewable energy. *Robotica, 38*(9), 1576–1640.

Hermawan, E., & Sudjadji, U. (2022). Integrated nuclear-renewable energy system for industrialisation in West Nusa Tonggara Province, Indonesia: Economic, potential site and policy recommendations. *International Journal of Energy Economics and Policy, 12*(4), 146–159. https://doi.org/10.32479/ijeep.13067

Issa, M., Ilinca, A., Rousse, D. R.,Boulon, L., & Groleau, P. (2023). Renewable energy and decarbonisation in the Canadian mining industry: Opportunities and challenges. *Energies, 16*(19). https://doi.org/10.3390/en16196967

Kaggwa, M., Mutanga, S., & Simelane, T. (2011). Factors determining the affordability of renewable energy—A note for South Africa. Policy brief No 65, AISA, Pretoria.

Kahia, M., Omri, A., & Jarraya, B. (2021). Does green energy complement economic growth for achieving environmental sustainability? Evidence from Saudi Arabia. *Sustainability, 13*, 180. https://doi.org/10.3390/su13010180

Lombard, A., & Ferreira, S. (2015). The spatial distribution of renewable energy infrastructure in three particular provinces of South Africa. https://doi.org/10.1515/bog-2015-0036

Ministry of Energy of the Russian Federation. (2017). The BRICS countries agreed to increase cooperation in the field of energy efficiency. Available at http://minergo.gov.ru/node/5571. Accessed 18 January, 2024.

Nape, K. M. et al. (2019). Introduction of household biogas digesters in rural farming households of the Maluti-a-Phofung municipality, South Africa. https://doi.org/10.17159/2413-3051/2019/v30i2a5885

Ndlovu, V., & Inglesi-Lotz, R. (2019). Positioning South Africa's energy supply mix internationally. https://doi.org/10.17159/2413-3051/2019/v30i2a5409

Okorie, M., Inambao, F., Chiguvare, Z., & Alfeus, S. (2018). Renewable energy and African industrialisation: A participatory integrated approach in assessing concentrated solar potential in Naminbia. *International Journal of Mechanical Engineering and Technology., 9*, 509–524.

Pathak, L., & Shah, K. (2019). Renewable energy resources, policies and gaps in BRICS countries and the global impact. *Frontiers in Energy, 13*(3), 506–521.

Setiawan, A. B., & Anwar, M. S. (2022). The environmental Kuznets carve hypothesis: An empirical evidence in Indonesia. *Jejak-Jpurbal of Economics and Policy, 15*(1), 207–2020. https://doi.org/10.15294/jejak.v15i1.38084

Simelane, T. S., & Abdel-Rahman, M. (Eds.). (2011). *Energy transition in Africa*. Africa Institute of South Africa.

Simelane, T., & Knop, J. (2022). Chapter 5 Opportunities and options for energy cooperation among BRICS countries. In S. Zondi (Ed.), *The political Economy of Intro-BRICS Cooperation, International political economy series*. Palgrave-Macmillan. ISBN-10 3030973964. ISBN-13 978-303097396.

Simelane, T., Managa, R., Mutanga, S., & Check, N. A. (2023). In search for Scientific collaboration: South Africa's science diplomacy towards Africa In VenugopalanIttekkot, & Baweja, J. K. (Eds.), *Science, technology and innovation diplomacy in developing countries-perceptions and practice*. Springer. ISBN 978-981-19-6802-0

Simelane, T., & Inyang, H. (2023). Africa's science diplomacy. In: F. Onditi, McLarren, K., Ben-Nun, G., & Stivachtis, Y.A., Okoth, P. G. (Eds.), *The Palgrave Handbook of Diplomatic Thought and Practice in the Digital Age*. Palgrave Macmillan. ISBN 978-3-031-28213-3. https://doi.org/10.1007/978-3-031-28214-0

Tawalbeh, M., Asaad, S. M., Ali, A., Al Kindi, S. R., & Al-Othman, A. (2024). Developments of bioenergy. *Renewable Energy, 2*, 245–273. https://doi.org/10.1016/B978-0-323-95211-8.00006-3

Usman, M., & Balsalobre-Lorente, D. (2022). Environmental concern in the era of industrialisation: Can financial development, renewable energy and natural resources alleviate some load? *Energy Policy, 162*. https://doi.org/10.1016/j.enpol.2022.112780

Wang, D., Wang, L., & Li, R. (2022). Does the energy transition alleviate environmental degradation? Evidence from the high income, upper and lower income economies. *Energy Reviews, 44*, 1–11. https://doi.org/10.1016/j.esr.2022.100966

Wolde-Rufael, Y. (2009). Energy consumption and economic growth: The experience of African countries revisited. *Energy Economics, 31*(2), 217–224. https://doi.org/10.1016/j.eneco.2008.11.005

Zhang, M, Yang, Z, Lui, L., & Zhou, D. (2021). Impact of renewable energy investment on carbon emissions in China-an empirical study using a nonparametric additive regression model. *Science of the Total Environment, 785*. https://doi.org/10.1016/j.scitotenv.2021.147109

Chapter 13
Re-industrialisation of De-industrialised Textile Sectors in Zimbabwe and Nigeria: Towards Achieving Sustainable Development Goal (SDG)-9

Idowu Jamiu Diyaolu and Kikelomo Olukemi Adubi

Abstract The textile sector is vital, as it is responsible for the production of materials used in clothing, furnishing, construction, medical and automobile industries. The sector has an interconnected production chain and has been a key driver of the industrial revolution. There are evidences that the textile sector previously contributed significantly to the economies of Zimbabwe and Nigeria, through increased GDP, employment generation, and foreign direct investment. However, both countries are presently de-industrialised, as they rely heavily on imported fabrics. There is an influx of cheap textiles, high production costs and lack of export initiatives, poor management and limited access to raw materials in both countries. This study adopted a review based methodology and answered questions such as: How can the textile firms in Zimbabwe and Nigeria be re-industrialised? Are there technological frameworks that could sustain the industry? The study examined innovative strategies to achieve Sustainable Development Goal 9 (SDG 9), as embraced in Industry 4.0. A technology framework for the re-industrialisation of the textile sector was proposed. The study identified the use of eco-friendly materials, the adoption of Industry 4.0, digital marketing and increased budgetary allocation to the textile sector as key steps toward re-industrialisation. This will drive other sectors, such as chemical, petrochemical, agricultural and other allied industries. The need for sustainability in material selection, eco-design, packaging and supply chain management, under an enabling policy environment, was also identified. The study recommends that the governments of Zimbabwe and Nigeria should adopt Industry 4.0, digital marketing and green production with environmentally friendly alternatives to revitalise and create a competitive niche in textile manufacturing.

I. J. Diyaolu (✉)
Department of Family, Nutrition and Consumer Sciences, Faculty of Agriculture, Obafemi Awolowo University, Ile-Ife, Nigeria
e-mail: diyaolu@oauife.edu.ng

K. O. Adubi
Federal University of Agriculture, Abeokuta, Nigeria

X. Poshiwa et al. (eds.), *Technological Innovations for Sustainable Industrialization*,
https://doi.org/10.1007/978-981-96-7995-9_13

Keywords Textile and clothing · Industrialisation · Zimbabwe and Nigeria · Sustainability

Introduction

Textiles and clothing (TC) are essential for human existence. There are various applications of TC in everyday life. They are used for protection and covering the body. The automobile industry uses textile products extensively in car upholstery, tyres and seat belt. Today, space travel would be impossible without smart textiles that are specially designed for the purpose. Hospitals make use of bandages, plasters and suture fibres during surgical operations. Geo-textiles and technical textiles are used in building and road construction.

Thus, the application of textiles to humans, animals and plants cannot be quantified. It is estimated that the total clothing sales will reach 160 million tonnes by 2050 (Ellen MacArthur Foundation, 2017).

The textile industry is responsible for the production of textiles and clothing. Textile industries in Zimbabwe and Nigeria were actively productive in the past. They contributed substantially to GDP, employment generation and exports to other nations but are recently folding up.

The textile manufacturing industry in Zimbabwe has faced many challenges and is currently operating at 30% below its full capacity. The industry that once employed over 40,000 people in its peak time now employs only 8000 workers (Bizvibe, 2022).

A similar observation is reported in the Nigerian textiles and clothing industry. Between 2003 and 2008, the number of active textile mills declined from 60 to 45, with a sharp drop in direct employment from 60,000 to 24,000. During this period, there was a steep fall in cotton lint production from about 90,000 tonnes to 60,000 tonnes. The export of textiles also declined. Many countries in Southwest Asia have used technological spillovers from textiles and apparel-related Foreign Direct Investment (FDI) to industrialise and diversify their economies, as well as integrate their markets into the global economy (Adejuwon & Oladale, 2019).

Few Sub-Saharan-based countries have benefited from the African Growth and Opportunity Act (AGOA). Industrial policy and weak industrialisation contributed to a de-industrialised economy (Rodrik, 2016).

However, developing countries can greatly benefit from cleaner technologies for sustainable textile production by facilitating technology transfer from developed countries through exchanges and collaborations (Adejuwon & Oladale, 2019; Diyaolu, 2022). Industrial upgrading could restore competitiveness in the textiles and clothing industry (Mvula & Chummun, 2020). The inclusion of brand innovation, supply chain management and sustainable design and manufacturing in the existing programs in Zimbabwe is suggested (Chuma, 2022).

The Focus of the Chapter

The study employed a review methodology on relevant literature, journals and online documents to update discussions on the industrialisation of the textile sector. The search string consisted of clothing, textiles, industrialisation, Zimbabwe, Nigeria, and sustainable development.

First, the study establishes a relationship between textile production and economic development by comparing the economic indicators of Zimbabwe and Nigeria with Switzerland. Switzerland was the most innovative economy in the world in 2022, followed by the USA, Sweden, the United Kingdom and the Netherlands (Dutta et al., 2022).

Secondly, the process of industrialisation and the causes of de-industrialization were examined. The study develops a conceptual framework and addresses relevant issues in the textile value chains. Questions relating to the current state of textile production in Zimbabwe and Nigeria, factors that contributed to the closure of the textile firms, and strategies to re-industrialise them were explored.

Sub-Saharan African countries are among the lowest-income countries in the world despite their significant endowment of natural resources (Yulek & Yagmur, 2018). Kaldor (1967) hypothesised that manufacturing sector is the engine of economic growth.

The economic indicators of Zimbabwe and Nigeria provide insight into industrialisation in the two countries. To some extent, Zimbabwe and Nigeria have comparable Global Innovation Index (GII) values in the 2022 rankings.

The output and input ranks, income and region classifications share several similar GII variables, as revealed in Table 13.1. The Global Innovation Index (GII) tracks the most recent global innovation trends against the backdrop of an ongoing COVID-19 pandemic, slowing productivity and other evolving challenges (WIPO, 2022).

Switzerland is the leading manufacturer of textile machinery worldwide, with an estimated global market share of 40% (Fibre2Fashion, 2022). A typical example

Table 13.1 Global innovation index (GII) of Zimbabwe and Nigeria (compared with Switzerland)

GII variables	Zimbabwe	Nigeria	Switzerland
GII 2022 rank	107	114	1
Output rank	93	107	1
Input rank	120	113	3
Income	Lower middle	Lower middle	High
Region	SSA	SSA	EUR
Population (mn)	15.1	211.4	8.7
GDP, PPP$ (bn)	37.3	1,136.8	677.3
GDP per capita, PPP$	2408	5377	78,112

Source Adapted from Dutta et al. (2022)

is Schoeller Textil AG, in Switzerland, which specialises in the sustainable development and production of innovative textiles (Schoeller Switzerland, 2022). If the governments of Zimbabwe and Nigeria were to emulate this sustainable approach, their economies could bounce back in textile manufacturing.

Challenges Facing the Textile Industry

The smuggling of imported textiles, a lack of adequate power supply and low-profit margins are prominent factors affecting textile production in developing countries (Diyaolu, 2017). Significant quantities of textiles and apparel are illegally imported into Zimbabwe and Nigeria.

About 85% of the textile materials in Nigeria are smuggled into the country. Akarakiri et al. (2012) argued that an agreement signed in November 2004 between Nigeria and the Republic of Benin, which allowed free passage of goods across their common borders, compounded the menace of smuggling.

The TC manufacturing process is highly energy-intensive, accounting for between 30 and 40% of the production costs. The insufficient supply of power to textile firms has significantly raised operating costs. According to a report from the Federal Ministry of Industry, Trade and Investment (2014), other challenges facing the TC industry include:

(i) Raw materials: The industry depends on imports of raw materials. These include a significant proportion of cotton and almost all dyes and chemicals. The reliance on imported raw materials and the problem of sourcing foreign exchange have contributed to the high cost of production.
(ii) Quality: The quality of cotton supplied to many ginneries is low due to contamination. This ultimately affects the quality of products.
(iii) Old textile plants and equipment: Most textile companies use old plants and outdated technology (Diyaolu, 2017). This also adds to the high costs of operations.

Sakoinakon et al. (2009) observed other impediments, such as poor technical know-how, weak supply chains, and ineffective marketing strategies in textile firm.

De-industrialisation Experience in Zimbabwe and Nigeria

De-industrialisation is a decline of industrial activity or capacity in a region or economy. Although the COVID-19 pandemic contributed to de-industrialisation in the global TC industry, it has also demonstrated the importance of technological innovation and resilient infrastructure in achieving re-industrialisation and Sustainable Development Goals (SDGs).

In 2021, global manufacturing rebounded from the pandemic, although the recovery remained incomplete and uneven. In developing countries, recovery has been sluggish and remains uncertain; almost one in three manufacturing jobs was negatively impacted by the crisis (United Nations, 2022).

Industrialisation policies have influenced manufacturing in Zimbabwe and Nigeria. A general import substitution policy is an example that was used in East Asian Countries. South Korean export promotion policies were employed simultaneously with import substitution (Yulek, 2016). Other factors that could lead to a de-industrialised TC in Zimbabwe and Nigeria include weak infrastructure that drives up the costs of production, poor transport systems, low internet connectivity and inadequate utility services.

Industry 4.0 and Sustainable Textile

Industrialization is the transformation from an agricultural resource-based economy to mass manufacturing, with increased income and improved living standards. Some significant studies on Sub-Saharan Africa TC industrialisation have been conducted (Yülek, 2016, 2017; Yülek et al., 2019a, 2019b; Yülek, & Yagmur, 2018). Findings from the study are applicable to the experiences of Zimbabwe and Nigeria. The countries can make use of the global textile and clothing market to transform their production and export.

Industry 4.0 describes the future of industry as the establishment of the internet and technological information based on interactive platforms (Cheng et al., 2016). It accelerates the industrial transition towards circularity (Happonen & Ghoreishi, 2021). Digital technologies help transfer real-time information on material and product condition, availability, and accessibility and boost the transition in the textiles and apparel industries. Product design development, product prototyping and recycling of materials can be done with higher efficiency by utilizing industry 4.0 technologies (Sinisterra et al., 2017).

Digital design and manufacturing have given the TC industry new opportunities for mass customisation (Kortelainen et al., 2019). Table 13.2 indicates the various stages of the industrial revolution, highlighting the progression from the first industrial revolution to the fourth industrial revolution. Industry 4.0 can promote textile firms in Zimbabwe and Nigeria by enhancing their sustainability.

It leads to virtualisation, modularisation/optimisation of production and the supply chain, achieving flexibility as well as personalisation of production (Brettel et al., 2014; Happonen & Minashkina, 2019).

Table 13.2 Development of industrial revolutions

Industrial revolution	Period and main feature
First industrial revolution	Began in the middle of the eighteenth century when the steam engine achieved the mechanisation of the factory
Second industrial revolution	Started in the mid-nineteenth century, with the power to implement mass production on a large scale
Third industrial revolution	Occurred in the mid-twentieth century through electrical and information technology to get automated manufacturing
Fourth industrial revolution	Based on the later evolution of the industry, according with physical system of cybernetics to develop new manufacturing methods

Source Sinisterra et al., 2017

Incorporating Sustainability in the Textile Supply Chain

A clothing and textile firm is a multi-sectoral industry connected with other agro-allied companies. The production of raw materials like cotton, flax and jute (plant fibres) starts on the farm. This is similar to wool and hair (animal fibres), which are obtained from animals.

Other fibres are synthetic, derived from oil, coal and other sources. They undergo various processes such as ginning, spinning, weaving, dyeing and finishing. Chemical industries that produce mordants, dyes, caustic soda and other fabric finishes are involved. This demonstrates that the textile industry is connected with other firms (Fig. 13.1).

In addition, textile manufacturing makes use of automated machines. A high level of technology in production is required for successful manufacturing. The ginning machine, carding, spinning, weaving, dyeing and finishing sections are all mechanised. This also implies that textile manufacturing is labour-intensive, and is considered a high employment-generating industry.

How Re-industrialisation Can Be Achieved in the Textile Industry

Re-industrialisation implies the economic, social and political process of organising national resources for re-establishing industries. To achieve re-industrialisation, textile production should incorporate regenerative and restorative agricultural practices that utilise organic inputs to retain and restore nutrients in the soil when growing cotton. Renewable energy sources could foster green textile industries (African Circular Economy Alliance, 2022).

Textile production in Kadoma, Zimbabwe needs sustainable approaches and should incorporate environmental, social and economic sustainability (Rugedhla

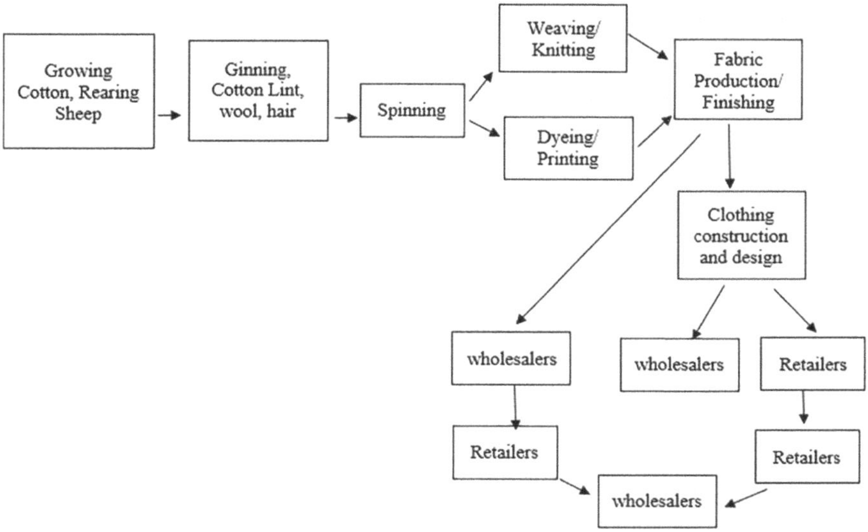

Fig. 13.1 Textile supply chain. *Source* Adapted from (McNamara, 2008; Diyaolu, 2016)

et al., 2020). Sustainable practices could help developing countries improve their production capacity (Diyaolu, 2021).

Srabani (2021) documented the need for sustainability in the textile industry, emphasising the use of natural resources instead of the chemicals commonly used in the textile production. The use of chemicals has led to an increase in water pollution.

Textiles production contributes to global warming due to the high consumption of fossil fuels consumed wet processing. Fossil fuels contain large amount of carbon and react with oxygen to form carbon dioxide, which increases greenhouse gas emissions and contributes to global warming. The following are identified steps towards achieving sustainable production in textiles:

i. **Utilisation of Eco-friendly Materials and Eco-design**

Sustainable textiles start with sustainable materials that come from eco-friendly resources like rayon, viscose and natural materials (Srabani, 2021). Natural materials tend to be much more sustainable. Some of the most popular natural materials used in the garment industry are cotton, wool, linen, hemp, mulberry and ramie. Organic fabrics are recyclable and biodegradable. The eco-design approach takes into account all possible factors such as ecological impact, recyclability, ergonomics and environmental friendliness. The textile firms in Zimbabwe and Nigeria should design products that cause less environmental damage while also increasing the aesthetic and functional properties of their products.

ii. **Selection of Non-toxic Alternatives in Production**

Environmentally friendly alternatives should be adopted during the textile manufacturing. These can include:

- Whitening (water-saving air dyeing, laser whitening, ozone whitening and finishing).
- Printing (digital printing technology).
- Mercerisation (electrochemical cell mercerisation).
- Waste-free fashion design, seamless knitting technology and integrated 3D design technology.
- Alternate wet process methods (chlorine-free bleaching, cold or low-temperature dye processes, dye bath reuse, eco bleach, Ink-jet printing, vegetable tanning, wastewater recycling) (Srabani, 2021).

iii. **Reduce-Reuse-Recycle Approach to Packaging**

Sustainable packaging implies packaging that incorporates the three principles of sustainability: reduce, reuse, and recycle. The main goal of sustainable packaging is to do as little harm to the environment as possible.

This can be achieved by using recycled materials or creating special packaging functionality (Srabani, 2021). Commercial recycling industries that recycle clothes back into yarns or upcycle secondhand clothing to added-value fabrics and other products will promote textile firms in Zimbabwe and Nigeria.

iv. **Transportation**

Transportation in the textile industry may take place through air, water, sea, rail or road. The transportation process consumes energy, leading to emissions of carbon dioxide, which in turn, negatively affects the environment. Poor roads and congested ports also increase the cost of moving raw materials around and shipping out finished goods. In addition, textile firms must focus on sustainability in the textile industry by improving resource productivity, co-efficiency, cost efficiency, customer satisfaction, brand reputation, environmental conditions and ensuring better health conditions for both wearers and workers.

v. **Availability of Infrastructure and Export Processing Zone (EPZ)**

The construction of EPZ or industrial zones serves as an advantage for AGOA. This means that firms could easily relocate to an AGOA benefiting country which already has infrastructure and facilities in place (Shakya, 2011).

Adejuwon and Oladale (2019) recommended the establishment of local firms, situating them in proximity to foreign firms, as well as the skill upgrading of the local labour force, especially in upstream value chain functions. Adu-Gyamfi (2020) proposed that to promote sustainable development in these EPZs, policy action should concentrate on attracting firms that are willing and able to align with the particular Sustainable Development Goal that zone programmes are intended to achieve. Cotton, Textile and Garment (CTG) infrastructure can ease the TC industry activities. Integrated Textiles and Garments Parks (ITGPs) should be developed close to raw materials and near the market. The ITGP will furnish textile firms with world-class facilities needed for setting up competitive units (Federal Ministry of Industry, Trade and Investment, 2015).

Framework for Enhanced Performance of the TC Industry in Zimbabwe and Nigeria

The proposed structural model to assist the textile manufacturing industry in Zimbabwe and Nigeria indicates that there is a relationship between industrial upgrading, competitiveness and business sustainability (Mvula & Chummun, 2020). Industrial upgrading is a strategy that improves the performance of the organisation. Product differentiation, low cost and continuous innovation were strategies that kept existing TC firms in operation (Diyaolu, 2017) (Fig. 13.2).

The textile industry is in dire need of a workable, sustainable and effective policy framework. While efforts have been made over the years, little has thus far been achieved to empower the industry. A comprehensive analysis should include fiscal policy, industrial and sectoral policy, investment promotion, export promotion and trade agreements.

Over the years, a major constraint has been the implementation of this policy (Ogah, 2015). One of the issues in the policy is the development of integrated textiles and garment parks (ITGPS) near raw materials, markets and requisite infrastructure, as achieved in other countries with thriving textiles sectors.

The government should, therefore, give attention to the textile parks. Diyaolu et al. (2018) and Diyaolu (2022) suggested that technology can stimulate improved productivity, improved quality, reduced processing costs, reduced processing time, increased production capacity, reduced environmental pollution and increased customer loyalty (Fig. 13.3).

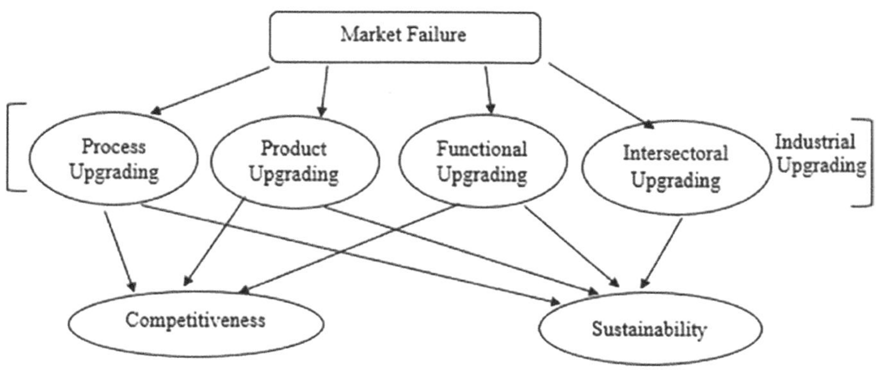

Fig. 13.2 Proposed model for upgrading the textile industry. *Source* Mvula & Chummun (2020)

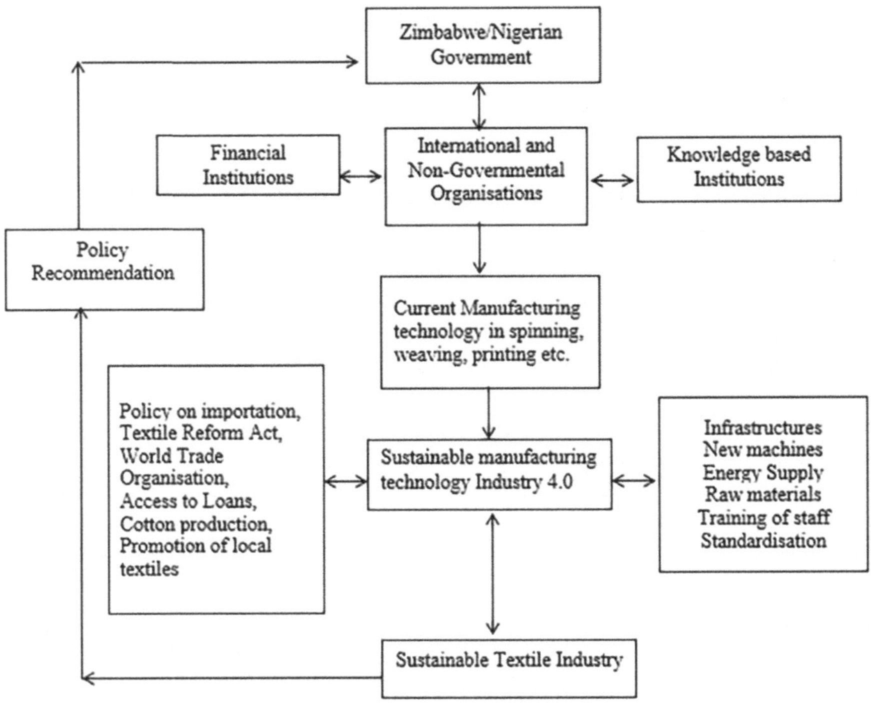

Fig. 13.3 Framework to promote and improve textile production in Zimbabwe and Nigeria. *Source* Adapted from Diyaolu (2016)

Role of Knowledge-Based Institution: There should be a university-industry partnership where research and development (R&D) in textile technology, innovation and design can be transferred. Challenges facing production in the industry should also be addressed by seeking solutions from institutions. Poor collaboration between universities and industry will hinder the TC industry from being re-industrialised.

Policy Issues: Policy related to TC manufacturing in Zimbabwe and Nigeria should be critically examined to foster production, patronage and use of local fabrics among consumers as well as governmental and non-governmental organisations. A notable factor hindering the firms is the lack of patronage of local products at the expense of imported ones. Policies surrounding AGOA, WTO and other organisations should be implemented to favour TC firms.

Such policies should favour raw material acquisition. Growing some of these raw materials will generate employment and cut costs of the import of fibres, dyes or other chemicals. Policy on sustainable products should be enacted in Nigeria and Zimbabwe.

This will go a long way towards promoting the use of local materials, reduce toxic materials and enabling a sustainable environment. Toxic textile waste, effluents and landfills can be sustainably converted into other useful products. Industry 4.0 should be embraced.

Support from Financial Institutions: There is no doubt that the TC re-industrialisation will be capital-intensive. Since we cannot do without clothing materials, budgetary allocation for the TC industry should be increased.

Financial institutions should offer loans to moribund TC firms to revive them. This can extend to the provision of good roads, energy, electricity and internet facilities.

Standardisation: Standardisation is equally important to a viable textile firm. TC products should conform to standard that do not affect human health. Training and retraining of staff will assist in the dissemination of new technology to production staff.

Conclusion and Recommendations

Textile firms in Zimbabwe and Nigeria were industrialised in the past. Due to the influx of foreign textiles into the countries and poor policy enactment in the sector, de-industrialisation set in. The two nations still have prospects for re-industrialisation, especially with the application of Industry 4.0, the sourcing of bio-degradable materials and among others.

The TC industry is very significant to the development of many nations. The social, economic and other aspects of a nation are strongly dictated by the state of its textile industry. That is why Zimbabwe and Nigeria need to pay close attention to the current state of their textile firms.

The study recommends:

i. Adopting industry 4.0 and modern techniques for TC production;
ii. utilisation of eco-friendly materials and eco-friendly environmental strategies;
iii. creating an export processing zone for TC production;
iv. training and retraining of staff; and
v. ensuring standardisation in every value chain.

References

Adejuwon, O. O., & Oladele, O. O. (2019). The geese that fail to fly: The AGOA, textile and apparel related FDI and implications for technological spillovers in Sub-Saharan Africa. In J. B. Akarakiri (Ed.), *Readings in technology management* (pp. 1–23). Obafemi Awolowo University Press.

Adu-Gyamfi, R., Asongu, S. A., Mmusi, T. S., Wamalwa, H., & Mangori, M. (2020). A comparative study of export processing zones in the wake of sustainable development goals: Cases of Botswana, Kenya, Tanzania and Zimbabwe. Research Africa Network (RAN) Working Paper WP/20/025.

African Circular Economy Alliance. (2022). Five Big Bets for Africa's path to circularity. African Circular Economy Alliance - Final Infographics.pdf

Akarakiri, J. B., Irefin, I. A., & Olateju, O. I. (2012). An evaluation of the effects of project management techniques on the performance of the textile industry in South-Western Nigeria. *Transnational Journal of Science and Technology, 2*(4), 28–35.

Bizvibe. (2022). Zimbabwe apparel industry struggles despite apparel exports increase (bizvibe.com). https://blog.bizvibe.com/blog/textiles-and-garments/zimbabwes-apparel-industry-struggles#:~:text=The%20apparel%20sector%20in%20Zimbabwe%20currently%20operates%20at,in%20peak%20time%20now%20employs%20only%208%2C000%20workers

Brettel, M., Friederichsen, N., Keller, M., & Rosenberg, M. (2014). How virtualization, decentralization and network building change the manufacturing landscape: An Industry 4.0 perspective. *International Journal of Industrial Engineering Computations, 8*(1), 1–8.

Cheng, G., Liu, L., Qianq, X., Liu, Y. (2016). Industry 4.0 development and application of intelligent manufacturing. In *International conference on information system and artificial intelligence*, Hong Kong.

Chuma, C. (2022). Curriculum review and education 5.0 in higher education in Zimbabwe: A review of textiles and clothing-related programs. *The Chitrolekha Journal on Art and Design, 6*(1), 1–9. https://doi.org/10.21659/cjad.61.v6n103

Diyaolu, I. J. (2016). Technology assessment of Nigerian textile industry. Ph.D. Thesis. African Institute for Science Policy and Innovation, Obafemi Awolowo University, Ile-Ife, Nigeria.

Diyaolu, I. J. (2022). Total quality management practices in nigerian textile industry. In: S. Msahli, Debbabi, F. (Eds.) *Advances in Applied Research on Textile and Materials - IX. CIRATM 2020* (Vol. 17). Springer Proceedings in Materials book series. Springer, Cham. https://doi.org/10.1007/978-3-031-08842-1_39

Diyaolu, I. J. (2017). Factors influencing textile production in Nigeria. *Journal of Home Economics Research, 24*(2), 93–100.

Diyaolu, I. J. (2021). Adoption of sustainability in clothing and textile production among developing countries. *Journal of Environment and Sustainable Development, 1*(1), 49–57. https://doi.org/10.55921/WFZA5978

Diyaolu, I. J., Irefin, I. A., & Akarakiri, J. B. (2018). Technology choices in Nigerian textile industry. *International Journal of Home Economics, 11*(1), 3–8.

Dutta, S., Lanvin, B., Wunsch-Vincent, S., & León, L. R. (2022). *Global innovation index 2022: What is the future of innovation-driven growth?* (15th ed, p. 266) ISBN: 9789280534337 online, https://doi.org/10.34667/tind.46596

Ellen MacArthur Foundation. (2017). A new textiles economy: Redesigning fashion's future. http://www.ellenmacarthurfoundation.org/publications

Federal Ministry of Industry, Trade and Investment. (2015). National cotton, textile and garment policy: A framework for the sub-sector. Federal Ministry of Industry, Trade and Investment, Federal Republic of Nigeria.

Fibre2Fashion. (2022). https://www.fibre2fashion.com/industry-article/60/switzerland-leading-textile-machinery-manufacturer-in-the-world#:~:text=The%20Swiss%20play%20a%20leading%20role%20in%20the,economies%2C%20Switzerland%20gained%20economic%20importance%20in%20the%20world

Happonen, A., & Ghoreishi, M. (2021). A mapping study of the current literature on digitalization and industry 4.0 technologies utilization for sustainability and circular economy in textile industries. Available at: https://www.researchgate.net/publication/348886083. https://doi.org/10.1007/978-981-16-2102-4_63

Happonen, A., & Minashkina, D. (2019). Operations automatization and digitalization—A research and innovation collaboration in physical warehousing, improvement ideas in AS/RS and 3PL logistics context. *LUT Research Reports Series Report, 104*, 1–41. https://doi.org/10.5281/zenodo.38137748

Kaldor, N. (1967). *Strategic factors in economic development*. New York State School of Industry and Labour Relations, Cornell University.

Kortelainen, H., Happonen, A., & Hanski, J., (2019). From asset provider to knowledge company—transformation in the digital era. In *Lecture notes in mechanical engineering* (pp. 333–341). https://doi.org/10.1007/978-3-319-95711-1_33

McNamara, K. (2008). The global textile and garments industry: The role of information and communication technologies (ICTs) in exploiting the value chain. Information for Development Program, World Bank Group. www.infodev.org/innovation. Accessed 24 December, 2011.

Mvula, D., & Chummun, B. Z. (2020). Investigating industrial upgrading and business sustainability in the textile manufacturing industry in Zimbabwe. *Journal of Contemporary Management, 7*(2), 445–471.

Ogah, D. (2015). *Textile Manufacturers hinge industry's revival on new policy* (p. 14). Lagos, Nigeria.

Rodrik, D. (2016). Premature deindustrialization. *Journal of Economic Growth, 21*(1), 1–33.

Rugedhla, M., Disele, P. L. P., Moalosi, R., & Fidzani, L. C. (2020). Factors that affect sustainability in the textile design industry in Kadoma, Zimbabwe. In S. Muthu, M. Gardetti, (Eds.), *Sustainability in the textile and apparel industries. Sustainable textiles: Production, processing, manufacturing & chemistry*. Springer. https://doi.org/10.1007/978-3-030-37929-2_3

Sakoinakon, T. P., Aim-Im-Thain, S., & Khamanarong, S. (2009). Management strategy for administration of textile industries in a developing country: Case study of Thailand. *Journal of Business Case Studies, 5*(3), 1–5.

Shakya, M. (2011). Apparel export in Lesotho: The State's role in building critical mass for competitiveness, In P. Chuhan-Pole, & Angwafo, M. (Eds.) *Yes Africa can: Success stories from a dynamic continent* (p. 477). World Bank Publication.

Sinisterra, K. V. B., Mejía, S. M., & Molano, J. I. R. (2017). In J. C. Figueroa-García et al., (Eds.), *Industry 4.0 and its development in Colombian industry*. Springer International Publishing. WEA 2017, CCIS 742, pp. 312–323. https://doi.org/10.1007/978-3-319-66963-2_28

Srabani, M. (2021). The Sustainability in Textile Industry. Available at https://textilefocus.com/sustainability-textile-industry/

Schoeller Switzerland. (2022). https://www.schoeller-textiles.com/en/

United Nations. (2022). The sustainable development goals report 2022. Manufacturing growth, 2006–2021 (percentage) Available at: https://unstats.un.org/sdgs/report/2022/The-Sustainable-Development-Goals-Report-2022.pdf

WIPO. (2022). https://www.wipo.int/global_innovation_index/en/2022/

Yülek, M. A. (2017). On the middle-income trap, the industrialisation process and appropriate industrial policy. *Journal of Industry, Competition, and Trade, 17*(3), 325–348.

Yülek, M. A., & Yagmur, M. H. (2018). Textile and clothing sector, and the industrialisation of Sub-Saharan Africa. In U. R. Efobi, & Asongu, S. (Eds.), *Financing sustainable development in Africa*. https://doi.org/10.1007/978-3-319-78843-2_16

Yülek, M. A., Natsuda, K., Yağmur, M. H., & Akkemik, K. A. (2019a). *The textile and clothing cycle and industrialisation cycle. EconAnadolu 2015 Conference Paper*.

Yülek, M. A., Natsuda, K., Akkemik, K. A., & Yağmur, M. H. (2019b). The textile and clothing industrialization cycle. *Marmara Üniversitesi İktisadi ve İdari Bilimler Dergisi, 41*(1), 256–277, ISSN: 2149-1844. https://doi.org/10.14780/muiibd.582322

Yülek, M. A. (2016). Industrial policies and outcomes: A comparison of pre-liberalised Korean and Turkish Economies. *Journal of Comparative Asian Development, 15*(1), 179–200.

Chapter 14
Leveraging Raspberry Pi Microservers for Infrastructure Development and Sustainable Industrialisation

Riyad Dhuny, Arshad Ahmud Iqbal Peer, Nawaz Ali Mohamudally, and Geerish Suddul

Abstract This work outlines the potential of Raspberry Pi as a microserver to foster sustainable development in developing and least-developed countries. The Fourth Industrial Revolution has introduced enhancements like cloud computing, the Internet of Things and artificial intelligence, which, while potentially revolutionary, entail high costs, posing a barrier to use in these countries. This paper explores Raspberry Pi microservers as an affordable alternative to costly servers and hosting infrastructure, promoting industrialisation and infrastructure development. Several open-source software applications conducive to various sectors are discussed, demonstrating the adaptability of Raspberry Pi microservers. Raspberry Pi also finds use in an on-premise hosting setup, offering a cheaper substitute to cloud-based infrastructure. The number of virtual users a microserver can support is linked to the upload rate, suggesting that increasing upload capacity can enhance usage. In conclusion, Raspberry Pi microservers present an economical and versatile solution for fostering sustainable industrialisation and infrastructure development worldwide. Increased upload bandwidth are suggested to maximise this potential further.

Keywords Microserver · Raspberry Pi · IoT · Sustainable industrialisation · Cloud computing · Open-source software applications · Infrastructure development

R. Dhuny (✉) · A. A. I. Peer · N. A. Mohamudally · G. Suddul
University of Technology, Mauritius, La Tour Koenig, Pointe-aux-Sables, Port-Louis 11134, Mauritius
e-mail: dhuny@utm.ac.mu

A. A. I. Peer
e-mail: apeer@utm.ac.mu

N. A. Mohamudally
e-mail: alimohamudally@utm.ac.mu

G. Suddul
e-mail: g.suddul@utm.ac.mu

X. Poshiwa et al. (eds.), *Technological Innovations for Sustainable Industrialization*,
https://doi.org/10.1007/978-981-96-7995-9_14

Introduction

The Fourth Industrial Revolution (Industry 4.0) has introduced a new era of economic disruption, characterised by the integration of advanced technologies such as cloud computing, the Internet of Things (IoT) and artificial intelligence (AI), (Roodt & Koen, 2020). These advancements have the potential to revolutionise various industries and sectors, driving sustainable development and improving the overall quality of life (Ghobakhloo, 2020; Stock et al., 2018). According to the United Nations (UN) Sustainable Development Goal (SDG)-9, there is a significant focus on creating resilient infrastructure, promoting inclusive and sustainable industrialisation and fostering innovation.

However, costs associated with IT infrastructure, including cloud computing, can present a barrier to entry for start-up companies in developing and least-developed countries (Martens et al., 2012; Varnika, 2018). Another challenge is the lack of connectivity and adequate bandwidth in many parts of the world (Greengard, 2010). Among the targets for SDG 9 is item 9.c, which states, 'Significantly increased access to information and communications technology and strive to provide universal and affordable access to the Internet in least-developed countries by 2020' (United Nations, 2015).

A subset of increasing access to information also includes the ability to publish information and make applications online, for which web servers are required. This chapter explores the application of Raspberry Pi (RPI), specifically as microserver hardware hosting web servers.

This innovation presents an affordable, sustainable solution, subsequently promoting industrialisation and infrastructure development in various regions. However, any solution employing a microserver as a web server would depend on the internet infrastructure.

As a contribution to this area, this work aims to enlighten stakeholders on the ideal minimum bandwidth expected from Internet Service Providers (ISPs) to empower end users to use on-premise or home-hosted solutions for online visibility or access to their web applications.

Background and Context

Cloud computing has become an essential component of modern information technology (IT) infrastructure, offering a range of benefits such as scalability, flexibility and on-demand access to computing resources (Sasikala, 2011). While cloud computing allows companies to avoid the heavy costs associated with developing IT infrastructure in-house (Varnika, 2018), it still requires the companies to hire cloud computing services supported by good internet connectivity, which has financial implications (Makhlouf, 2020).

The financial constraints (Martens et al., 2012) can be incredibly restrictive for small enterprises, entrepreneurs and organisations in developing nations, hindering their capacity to utilise state-of-the-art technology for fostering innovation and expansion.

Among the uses of cloud computing is a model called infrastructure as a service (Firdhous, 2014; Sitaram & Manjunath, 2012), which offers virtual computing resources as a service that includes computing and storage.

Byrne et al. (2022) demonstrated the feasibility of replacing the existing server hardware within the data centres offering cloud services with ARM-based single-board computers (SBCs) to reduce the total cost of ownership, increase energy efficiency and offer virtualisation-free hardware access.

An SBC is defined by Jovanović et al. (2014) as '*a general-purpose computer built on a single board with microprocessor(s), memories, input/output(I/O) ports, and other features of a functional computer*'. SBCs are also used as part of solutions for the Internet of Things (IoT) and Industrial Internet of Things (IIoT) (Galkin et al., 2018). The Internet of Things (IoT) has surfaced as a vital element of Industry 4.0, covering a broad spectrum of interconnected devices, sensors and actuators that gather, process and share information (Griffiths & Ooi, 2018; Mohamed, 2019).Sensors and actuators can be connected to SBCs via I/O ports, as depicted in Fig. 14.1, which illustrates the Raspberry Pi version 4.

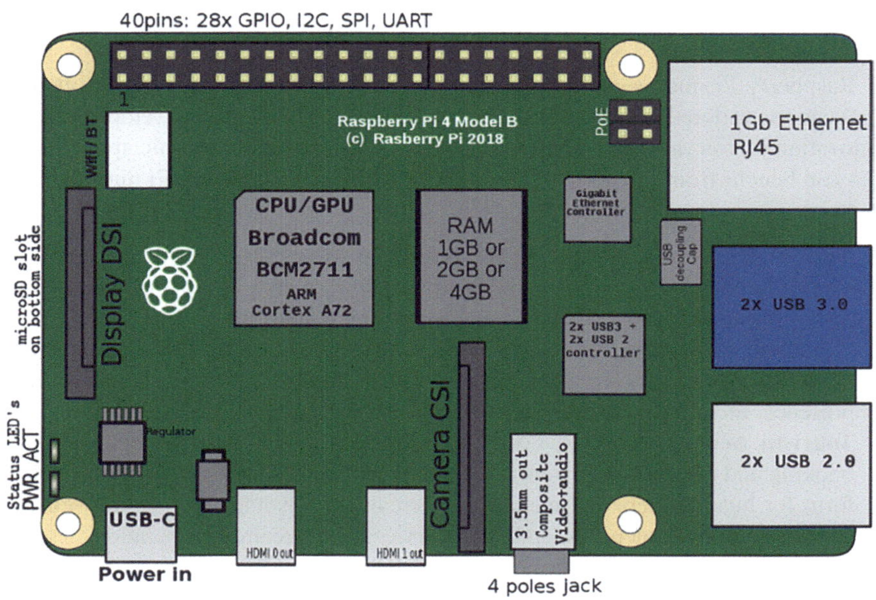

Fig. 14.1 Raspberry Pi 4 hardware drawing (Jstrom99, 2019)

The RPI is a credit-card-sized SBC that can be employed as an IoT hardware platform (Mohamed, 2019, Sect. 14.4) and a cost-effective, compact and adaptable computing solution for various purposes (Galkin et al., 2018).

With 46 Million units sold as of 2022 (Speed, 2022), it has become the third-world's best-selling computing platform in only a few years. As SBCs have evolved and become more potent over time, they have given birth to the notion of microservers, small, energy-efficient server systems capable of supporting an array of server applications (Anwar et al., 2014; Editya and Kurniati, 2020).

1. The Raspberry Pi Microserver Revolution: Leveraging Innovations for Infrastructure Development

The RPI (The MagPi, 2017) has gained popularity as a cost-effective and accessible computing platform, making it an attractive option for microserver applications. One notable example is the RPI64Box project (Dhuny, 2022; Dhuny et al., 2022), which transforms the RPI into a web, application and database server, offering a portable Linux, Apache, MySQL and PHP (LAMP) server setup (Louridas, 2016).

The RPI64Box converts the RPI's Wi-Fi into a hotspot, allowing other devices to connect and access applications hosted on the microserver. This portability and ease of use make Raspberry Pi microservers an ideal solution for knowledge sharing, education and disseminating open-source software applications in developing and least-developed countries.

2. Open-Source Applications for Sustainable Industrialisation

Raspberry Pi microservers can facilitate the adoption of open-source software applications tailored to specific sectors, promoting sustainable development and innovation across various industries. Some examples of sector-specific applications that can benefit from the portability and affordability of Raspberry Pi microservers include:

- **Agriculture**: Farm OS (Stenta, 2017) is an open-source, web-based farm management and record-keeping tool designed to help farmers plan, manage and track farming operations, livestock and environmental conditions.
- **Food Service Sector**: TastyIgniter (Adepoyigi, 2013) is an open-source management system that allows restaurant owners to establish and manage their online presence, including configurable menus, ordering and table reservations.
- **Tourism Sector**: QloApps (Webkul Software, 2016), an open-source hotel booking and reservation system built for the hospitality industry, provides a platform for hotel owners and managers to create and manage their hotel website, showcase rooms, amenities, and other services and streamline operations.
- **Customer Relationship Management** (CRM): SuiteCRM (SalesAgility, 2013) is an open-source CRM platform that offers sales force automation, marketing automation, customer support and collaboration tools.
- **Project Management**: phpCollab (Mindblender Productions, 2020) is an open-source project management and collaboration tool that helps teams coordinate and track projects, tasks and documents.

- **Human Resources Management** (HRM): OrangeHRM (Saparamadu, 2006) is a comprehensive open-source HRM system that helps businesses manage employee information, attendance, performance and other HR-related tasks.
- **Inventory Management**: PartKeepr (Hummel, 2010) is an inventory management system that assists organisations in tracking and managing electronic components, parts and other inventory products, thereby optimising inventory procedures and lowering the risk of stockouts or surplus inventory.
- **Education and Training**: Moodle (Dougiamas, 2002) is the most popular open-source e-learning platform for disseminating knowledge and classroom management.

3. Assessing Microserver Capabilities and Use Cases

Understanding the capabilities of Raspberry Pi microservers and their potential to support various applications and users is critical for evaluating their suitability for specific scenarios. Dhuny et al. (2022) experimented to determine the number of simultaneous users that could be supported by a Raspberry Pi version 4 running a Moodle application, the popular online learning platform.

The authors concluded that around 40 users could be easily supported with a medium-sized course on the 1.5 GHz CPU of RPI4 using the RPI64Box with a solid-state disk (SSD) as storage.

In a separate case study (Dhuny, 2021; Dhuny & Mohamudally, 2022) highlighting the versatility and applicability of Raspberry Pi microservers for outdoor usage and remote locations, the RPI64Box was used to support an archery competition, running the IanSeo (Scarzella et al., 2006) software, a score tracking system for the sport.

The club president and competition organiser praised the RPI64Box for its portability, built-in Wi-Fi hotspot and ability to quickly streamline the score review process and display results. The experimental test was an eye-opener for the organiser, leading to an immediate adoption of the technology.

4. On-Premise Hosting with Microservers

Raspberry Pi microservers can also be used for on-premise hosting of web applications and websites, offering an affordable alternative to cloud-based infrastructure. However, it is vital to know the user experience in terms of wait time.

The server's response time in on-premise hosting has among the dependent variables the internet bandwidth. Table 14.1 shows an ISP's upload and download speeds for a specific Internet package, with the first row representing the package label.

The experiment of Dhuny et al. (2022) evaluated the RPI4 as a microserver hardware acting as a web server. The tests could be repeated by limiting the bandwidth to specific Internet connection values, as in Table 14.1, to measure the impact of bandwidth limitation.

Table 14.1 Various internet connection speeds used during the test are representative of an ISP provider

Internet package label	20 Mbps	50 Mbps	100 Mbps	500 Mbps	1 Gbps
Download speed in Mbps	20	50	100	500	1000
Upload speed in Mbps	8	10	20	40	60
Max volume allowance in TB	1	2	4	10	15

Materials and Methods

As materials, the RPI64Box (Dhuny & Mohamudally, 2022), for which a disk image was available on its website (Dhuny, 2022), was used on the most recent release of RPI4 hardware with a CPU clock speed of 1.8 GHz (Upton, 2021) as a system under test (SUT).

A second Linux with Apache JMeter (Apache JMeter, 1998) was used as the driver to run a series of automated tests for performance evaluation. The test assumed that each user would take a thinking time of 7 s to go through the displayed contents before taking the next action.

The method and materials are similar to the study done by Dhuny et al. (2022), except that in this repeated experiment, the bandwidth is limited to represent the bandwidth provided by an ISP provider in Table 14.1. Both devices were connected to a 1Gbps Network router via the wired interfaces.

As a testbed, medium-sized course content from Moodle was used, and the number of users varied from 30 to 50. The Wondershaper tool (Bert et al., 2002) was used as the bandwidth limiter, and the 'iperf3' command (Dugan et al., 2014) was used to ensure the bandwidth was limited.

The results were measured using the Application Index (Apdex) value (Sevcik, 2005). Apdex is a range of values between 0 and 1, indicating the user experience waiting for a response from the webserver.

The Apdex value is rated as unacceptable if the value is below 0.5, and a value closer to 1 is considered excellent. Within Apdex, a user is considered to reach a frustration threshold if the wait time is several times the expected response time.

In the current case, a factor of three is used. Assuming a user's expected wait time for a reply is 0.5 s ($Apdex_{0.5}$). A user would be considered to reach the frustration threshold if a response is received after three times the expected value, i.e. 1.5 s. Within the Apdex unit of measure, such responses would bring the Apdex value below 0.5, for which the Apdex rating is considered unacceptable, as previously mentioned. Similarly, if the user's expectancy for a response from the server is raised to consider a reply within 1.0 s ($Apdex_{1.0}$), then the threshold value before frustration would be reached by 3.0 s. A scenario that can be envisaged in remote areas with poor internet connectivity to access information.

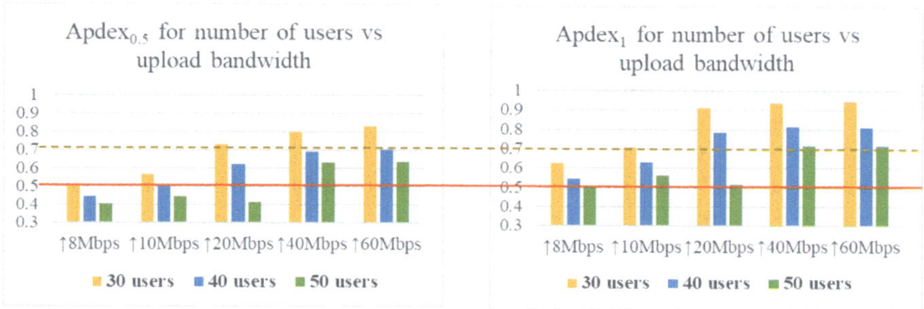

Fig. 14.2 Bar chart of increasing number of users against upload bandwidth for $Apdex_{0.5}$ (Left) and $Apdex_{1.0}$ (Right)

Results

The results demonstrated that a home-hosted or on-premise RPI4 could be used with an Internet connectivity upload speed of at least 8Mbps. The microserver's ability to support more virtual users improved as the upload rate increased. For the end-user wait time experience, the results were measured using an Application Index (Apdex) unit, which ranges from 0 to 1, as displayed on the vertical axis of Fig. 14.2.

Assuming our end users have high exigencies and expect an answer within half a second, the outcome would be that of Fig. 14.2Left. It shows that 30 users would have an experience just above the 0.5-mark, considered poor but acceptable on an 8Mbps upload line.

The microserver could support up to 50 virtual users when end-users considered a one-second wait time to be an excellent response time, as shown in Fig. 14.2Right.

In Fig. 14.2, the 0.7-mark line represents an Apdex rating considered fair by the end users. This value indicates the number of users a microserver can support in a production environment. It is a valuable indicator representing a fair user experience regarding wait time.

Conclusion

Microservers based on Raspberry Pi have surfaced as robust and economical options for operating various server applications throughout multiple sectors and industries. The extensive accessibility of open-source software allows businesses and institutions in developing and underdeveloped nations to explore and utilise advanced technologies without substantial upfront expenses in cloud infrastructure. Employing Raspberry Pi microservers like the RPI64Box, aids in fostering education, knowledge dissemination and the implementation of open-source software applications designed for particular sectors. This approach encourages sustainable growth and creativity, propelling infrastructure advancement and industrialisation in developing and least-developed countries. The RPI64Box is freely available for download at rpi 64box.com. The default web application can be easily replaced with new ones and is an excellent way to get started. As Raspberry Pi microservers evolve and become more powerful, their potential to support a growing range of applications and users will increase. This evolution will further expand their applicability and value for promoting sustainable industrialisation and infrastructure development worldwide.

However, one limitation remains: the bandwidth. ISPs can help with SDG target 9.c by offering higher upload bandwidth in areas with asymmetric upload and download speeds. An upload value of 20Mbps would be an ideal starting point for optimal support of approximately 30 users in a home-hosted setup.

Support from ISPs would enable end users to have their own cloud networks using affordably priced microservers. This action can be an enabler for information access in developing countries.

References

Adepoyigi, S. (2013). About TastyIgniter–Restaurant online ordering system. Accessed April 6, 2023 Available at: https://tastyigniter.com

Anwar, A., Krish, K. R., & Butt, A. R. (2014). On the use of microservers in supporting Hadoop applications. In *2014 IEEE international conference on cluster computing (CLUSTER)*, pp. 66–74.

Apache JMeter. (1998). Apache JMeter. Accessed May 30, 2021 Available at: https://jmeter.apache.org/

Bert, H., Jacco, G., & Simon, S. (2002). Wondershaper. Accessed December 25, 2022 Available at: https://github.com/magnific0/wondershaper

Byrne, A., Pang, Y., Zou, A., Nadgowda, S., & Coskun, A. K. (2022). MicroFaaS: Energy-efficient Serverless on Bare-metal Single-board Computers. In *2022 Design, automation & test in Europe conference & exhibition (DATE)*, pp. 754–759.

Dhuny, R. (2021). RPI64Box-Ianseo. Accessed June 6, 2022. Available at: https://rpi64box.com/rpi64box-ianseo/

Dhuny, R. (2022). RPI64Box. Accessed July 19, 2022 Available at: https://rpi64box.com/

Dhuny, R., & Mohamudally, N. A. (2022). RPI64Box: A portable 3-tiered LAMP stack in a 64-bit operating system environment. *Software Impacts, 14*, Article 100390.

Dhuny, R., Peer, A. A. I., Mohamudally, N. A., & Nissanke, N. (2022). Performance evaluation of a portable single-board computer as a 3-tiered LAMP stack under 32-bit and 64-bit operating systems. *Array, 15*, Article 100196.

Dougiamas, M. (2002). Moodle. Accessed July 19, 2022 Available at: https://moodle.org/

Dugan, J., Elliott, S., Mah, B. A., Poskanzer, J., & Prabhu, K. (2014). iPerf–The TCP, UDP and SCTP network bandwidth measurement tool. Accessed April 6, 2021 Available at: https://iperf.fr/

Editya, A. S., & Kurniati, N. (2020). Microserver performance to handle heavy traffic in dedicated server host for local area network. In *2020 IEEE international conference on communication, networks and satellite (Comnetsat)*, pp. 170–173.

Firdhous, M. (2014). A comprehensive taxonomy for the infrastructure as a service in cloud computing. In *2014 fourth international conference on advances in computing and communications*, pp. 158–161.

Galkin, P., Golovkina, L., & Klyuchnyk, I. (2018). Analysis of single-board computers for IoT and IIoT solutions in embedded control systems. In *2018 international scientific-practical conference problems of Infocommunications. Science and technology (PIC S&T)*, pp. 297–302.

Ghobakhloo, M. (2020). Industry 4.0, digitisation, and opportunities for sustainability. *Journal of Cleaner Production, 252*, 119869.

Greengard, S. (2010). Cloud computing and developing nations. *Communication ACM, Association for Computing Machinery, New York, NY, USA, 53*(5), 18–20.

Griffiths, F., & Ooi, M. (2018). The fourth industrial revolution–Industry 4.0 and IoT [Trends in Future I&M]. *IEEE Instrumentation and Measurement Magazine, 21*(6), 29–43.

Hummel, F. (2010). PartKeepr–Open source inventory management. Accessed April 6, 2023 Available at: https://partkeepr.org/

Jovanović, P., Mileusnić, M., Pavić, B., & Mišković, B. (2014). Applications of the single board computers in the software defined radio systems. In *Sinteza 2014–Impact of the internet on business activities in serbia and worldwide*, pp. 882–886.

Jstrom99. (2019). RaspberryPi model 4B drawing. Accessed September 25, 2022 Available at: https://commons.wikimedia.org/wiki/File:RaspberryPi_Model_4B.svg

Louridas, P. (2016). Component stacks for enterprise applications. *IEEE Software, 33*(2), 93–98.

Makhlouf, R. (2020). Cloudy transaction costs: A dive into cloud computing economics. *Journal of Cloud Computing, 9*(1), 1.

Martens, B., Walterbusch, M., & Teuteberg, F. (2012). Costing of cloud computing services: A total cost of ownership approach. In *2012 45th Hawaii international conference on system sciences*, pp. 1563–1572.

Mindblender productions. (2020). phpCollab. Accessed July 18, 2023 Available at: https://www.phpcollab.com/

Mohamed, K. S. (2019). *IoT physical layer: Sensors, actuators, controllers and programming* (pp. 21–47). Towards a Smart World, Springer International Publishing, Cham.

Roodt, J. H., & Koen, H. (2020). A review of hurdles to adopting industry 4.0 in developing countries. *INCOSE International Symposium, 30*(1), 1079–1092.

SalesAgility. (2013). SuiteCRM–Open source CRM software application for businesses. Accessed July 18, 2023 Available at: https://suitecrm.com/

Saparamadu, S. (2006). OrangeHRM. Accessed July 18, 2023 Available at: https://www.orangehrm.com/

Sasikala, P. (2011). Cloud computing: Present status and future implications. *International Journal of Cloud Computing, 1*(1), 23–36.

Scarzella, A., Gabardi, A., Deligant, C., Pisani, M., & Carpignano, M. (2006). Ianseo. Accessed June 4, 2022 Available at: https://www.ianseo.net/

Sevcik, P. J. (2005). Defining the application performance index. *Business Communications Review, 20*, 8–10.

Sitaram, D., & Manjunath, G. (2012). Chapter 2–Infrastructure as a service. In D. Sitaram & G. Manjunath (Eds.), *Moving to the cloud* (pp. 23–71). Syngress.

Speed, R. (2022). One decade, 46 million units: Happy birthday, Raspberry Pi, *The Register*. Accessed July 17, 2025. https://www.theregister.com/2022/02/28/pi_at_10/

Stenta, M. (2017). farmOS|farmOS. Accessed April 6, 2023 Available at: https://farmos.org/

Stock, T., Obenaus, M., Kunz, S., & Kohl, H. (2018). Industry 4.0 as enabler for a sustainable development: A qualitative assessment of its ecological and social potential. *Process Safety and Environmental Protection, 118*, 254–267.

The MagPi. (2017). Sales soar above Commodore 64. *The MagPi*, No. 56, p. 8.

United Nations. (2015). *Transforming our world : The 2030 agenda for sustainable development: Draft resolution referred to the United Nations summit for the adoption of the post-2015 development agenda by the general assembly at its 69th session*, UN. Accessed June 21, 2023 Available at: https://digitallibrary.un.org/record/803352

Upton, E. (2021). Bullseye bonus: 1.8 GHz Raspberry Pi 4–Raspberry Pi. Accessed July 24, 2023 Available at: https://www.raspberrypi.com/news/bullseye-bonus-1-8ghz-raspberry-pi-4/

Varnika, V. S. (2018). Cloud computing advantages and challenges for developing nations. *Indian Journal of Computer Science, 3*(3), 36–39.

Webkul software. (2016). Open source and free hotel booking management software–QloApps. Accessed July 18, 2023 Available at: https://qloapps.com/

The manufacturer's authorised representative in the EU is Springer Nature Customer Service Centre GmbH, Europaplatz 3, 69115 Heidelberg, Germany. If you have any concerns regarding our products, please contact ProductSafety@springernature.com

Printed and bound by CPI Group (UK) Ltd, Croydon, CR0 4YY
21/09/2026
02220015-0005